Quality Assurance and Quality Control in the Analytical Chemical Laboratory

A Practical Approach
Second Edition

ANALYTICAL CHEMISTRY SERIES

Quality Assurance and Quality Control in the Analytical Chemical Laboratory

A Practical Approach
Second Edition

Piotr Konieczka
Jacek Namieśnik

CRC Press
Taylor & Francis Group
Boca Raton London New York

CRC Press is an imprint of the
Taylor & Francis Group, an **informa** business

CRC Press
Taylor & Francis Group
6000 Broken Sound Parkway NW, Suite 300
Boca Raton, FL 33487-2742

© 2018 by Taylor & Francis Group, LLC
CRC Press is an imprint of Taylor & Francis Group, an Informa business

No claim to original U.S. Government works

Printed on acid-free paper

International Standard Book Number-13: 978-1-138-19672-8 (Hardback)

Visit the Taylor & Francis Web site at
http://www.taylorandfrancis.com

and the CRC Press Web site at
http://www.crcpress.com

Contents

Preface

The aim of this book is to provide practical information about quality assurance/quality control (QA/QC) systems, including the definitions of all tools, an understanding of their uses, and an increase in knowledge about the practical application of statistical tools during analytical data treatment.

Although this book is primarily designed for students and teachers, it may also prove useful to the scientific community, particularly among those who are interested in QA/QC. With its comprehensive coverage, this book can be of particular interest to researchers in the industry and academia, as well as government agencies and legislative bodies.

The theoretical part of the book contains information on questions relating to quality control systems.

The practical part includes more than 80 examples relating to validation parameter measurements, using statistical tests, calculation of the margin of error, estimating uncertainty, and so on. For all examples, a constructed calculation datasheet (Excel) is attached, which makes problem solving easier.

The eResource files available to readers of this text contain more than 80 Excel datasheet files, each consisting of three main components: Problem, Data, and Solution. In some cases, additional data, such as graphs and conclusions, are also included. After saving an Excel file on the hard disk, it is possible to use it on different data sets. It should be noted that in order to obtain correct calculations, it is necessary to use it appropriately. The user's own data should be copied only into yellow marked cells (be sure that your data set fits the appropriate datasheet). Solution data will be calculated and can be read from green marked cells.

We hope that with this book, we can contribute to a better understanding of all problems connected with QA/QC.

About the Authors

Piotr Konieczka (MSc, 1989; PhD, 1994; DSc, 2008-GUT; Prof., 2014) has been employed at Gdańsk University of Technology since 1989 and is currently working as a full professor.

His published scientific output includes 2 books, 10 book chapters, and more than 80 papers published in international journals from the JCR list ($\sum$ IF = 188), as well as more than 100 lectures and communications. His number of citations (without self-citations) equals 957 and his h-index is 19 (according to the Web of Science, December 31, 2017). Dr. Konieczka has been supervisor or co-supervisor of six PhD theses (completed).

His research interests include metrology, environmental analytics and monitoring, and trace analysis.

Jacek Namieśnik (MSc, 1972-GUT; PhD, 1978-GUT; DSc, 1985-GUT; Prof., 1998) has been employed at Gdańsk University of Technology since 1972. Currently a full professor, he has also served as vice dean of the Chemical Faculty (1990–1996) and dean of the Chemical Faculty (1996–2000 and 2005–2012). He has been the head of the Department of Analytical Chemistry since 1995, as well as chairman of the Committee of Analytical Chemistry of the Polish Academy of Sciences since 2007, and Fellow of the International Union of Pure and Applied Chemistry (IUPAC) since 1996. He was director of the Centre of Excellence in Environmental Analysis and Monitoring in 2003–2005. Among his published scientific papers are 8 books, more than 700 papers published in international journals from the JCR list ($\sum$ IF = 1975), and more than 400 lectures and communications published in conference proceedings. Dr. Namieśnik has 10 patents to his name. His number of citations (without self-citations) equals 9415 and his h-index is 47 (according to Web of Science, October 31, 2017). He has been the supervisor or co-supervisor of 64 PhD theses (completed).

Dr. Namieśnik is the recipient of various awards, including Professor *honoris causa* from the University of Bucharest (Romania, 2000), the Jan Hevelius Scientific Award of Gdańsk City (2001), the Prime Minister of Republic of Poland Awards (2007, 2017), the award of the Ministry of Science and Higher Education Award for young scientists' education (2012), the award of the Ministry of Science and Higher Education Award for outstanding achievements in the field of society development (2015), and doctor honoris causa of the Military Technical Academy (Warsaw) and Medical University of Gdańsk (Gdańsk, 2015).

Dr. Namieśnik was elected as a rector of Gdańsk University of Technology for the period of 2016–2020. His research interests include environmental analytics and monitoring, and trace analysis.

List of Abbreviations

AAS	Atomic Absorption Spectrometry
ANOVA	Analysis Of Variance
BCR	Bureau Communautaire de Reference (Standards, Measurements, and Testing Programme–European Community)
CDF	Cumulative Distribution Function
CITAC	Cooperation on International Traceability in Analytical Chemistry
CL	Central Line
CRM	Certified Reference Material
CUSUM	Cumulative SUM
CV	Coefficient of Variation
CVAAS	Cold Vapor Atomic Absorption Spectrometry
D	Mean Absolute Deviation
EN	European Norm
GC	Gas Chromatography
GLP	Good Laboratory Practice
GUM	Guide to the Expression of Uncertainty in Measurement
IAEA	International Atomic Energy Agency
ICH	International Conference on Harmonisation
IDL	Instrumental Detection Limit
ILC	InterLaboratory Comparisons
IQC	Internal Quality Control
IQR	InterQuaRtile Value
IRMM	Institute for Reference Materials and Measurements
ISO	International Organization for Standardization
IUPAC	International Union of Pure and Applied Chemistry
JCGM	Joint Committee for Guides in Metrology
LAL	Lower Action (Control) Limit
LOD	Limit of Detection
LOQ	Limit of Quantification
L-PS	Laboratory-Performance Study
LRM	Laboratory Reference Material
LWL	Lower Warning Limit
MB	Method Blank
M-CS	Material-Certification Study
MDL	Method Detection Limit
Me	Median
Mo	Mode
M-PS	Method-Performance Study
MQL	Method Quantification Limit
NIES	National Institute for Environmental Studies
NIST	National Institute of Standards and Technology
NRCC	National Research Council of Canada

PRM	Primary Reference Material
PT	Proficiency Test
q	Quartile
QA$_{(1)}$	Quality Assessment
QA$_{(2)}$	Quality Assurance
QA/QC	Quality Assurance/Quality Control
QC	Quality Control
QCM	Quality Control Material
R	Range
RH	Relative Humidity
RM	Reference Material
RSD	Relative Standard Deviation
S/N	Signal-to-Noise Ratio
SD	Standard Deviation
SecRM	Secondary Reference Material
SI	Le Systeme Internationale d'Unités (The International System of Units)
SOP	Standard Operating Procedure
SRM	Standard Reference Material
UAL	Upper Action (Control) Limit
USP	The United States Pharmacopeia
UWL	Upper Warning Limit
V	Variance
VIM	Vocabulaire International des Termes Fondamentaux et Généraux de Métrologie (International Vocabulary of Metrology)
VIRM	The European Virtual Institute for Reference Materials
%R	Recovery

1 Basic Notions of Statistics

1.1 INTRODUCTION

Mathematical statistics is a branch of mathematics that applies the theory of probability to examining regularities in the occurrence of certain properties of material objects or phenomena which occur in unlimited quantities. Statistics presents these regularities by means of numbers.

Statistics is not only art for art's sake. It is a very useful tool that can help us find answers to many questions. Statistics is especially helpful for analysts, because it may clear many doubts and answer many questions associated with the nature of an analytic process, for example:

- How exact the result of determination is
- How many determinations should be conducted to increase the precision of a measurement
- Whether the investigated product fulfills the necessary requirements or norms

Yet it is important to remember that statistics should be applied in a reasonable way.

1.2 DISTRIBUTIONS OF RANDOM VARIABLES

1.2.1 CHARACTERIZATION OF DISTRIBUTIONS

The application of a certain analytical method unequivocally determines the distribution of measurement results (properties), here treated as independent random variables. A result is a consequence of a measurement. The set of obtained determination results creates a distribution (empirical).

Each defined distribution is characterized by the following parameters:

- A cumulative distribution function (CDF) X is determined by F_X and represents the probability that a random variable X takes on a value less than or equal to x; a CDF is (not necessarily strictly) right-continuous, with its limit equal to 1 for arguments approaching positive infinity, and equal to 0 for arguments approaching negative infinity; in practice, a CDF is described shortly by $F_X(x) = P(X \leq x)$.
- A density function which is the derivative of the CDF: $f(x) = F_X'(x)$.

Below are the short characterizations of the most frequently used distributions:

- Normal distribution
- Uniform distribution (rectangular)
- Triangular distribution

Normal distribution, also called *Gaussian distribution* (particularly in physics and engineering), is a very important probability distribution used in many domains. It is an infinite family of many distributions, defined by two parameters: mean (location) and standard deviation (scale).

Normal distribution, $N(\mu_x, SD)$, is characterized by the following properties:

- An expected value μ_x
- A median $Me = \mu_x$
- A variance V

Uniform distribution (also called *continuous* or *rectangular*) is a continuous probability distribution for which the probability density function within the interval $\langle -a, +a \rangle$ is constant and not equal to zero, but outside the interval is equal to zero.

Because this distribution is continuous, it is not important whether the endpoints $-a$ and $+a$ are included in the interval. The distribution is determined by a pair of parameters – a and $+a$.

Uniform distribution is characterized by

- An expected value $\mu_x = 0$
- A median $Me = 0$
- A variance $V = a^2/3$

Triangular distribution over the interval $\langle -a, +a \rangle$ is characterized by

- An expected value $\mu_x = 0$
- A median $Me = 0$
- A variance $V = a^2/6$

The distribution of a random variable provides complete information on an investigated characteristic (e.g., concentration, content, physiochemical property). Unfortunately, such complete information is seldom available. As a rule, characteristic inference is drawn using the analysis of a limited number of elements (samples) representing a fragment of the whole set that is described by the distribution. Then, one may infer a characteristic using an estimation of some of its parameters (statistical parameters) or its empirical distribution.

Statistical parameters are numerical quantities used in the systematic description of a statistical population structure.

These parameters can be divided into four basic groups:

- Measures of location
- Measures of statistical dispersion
- Measures of asymmetry
- Measures of concentration

1.3   MEASURES OF LOCATION

Measures of location use one value to characterize the general level of the value of the characteristic in a population [1].

The most popular measures of location are the following:

- Arithmetic mean
- Truncated mean
- Mode
- Quantiles
  - Quartiles
  - Median
  - Deciles

Arithmetic mean is the sum of all the values of a measurable characteristic divided by the number of units in a finite population:

$$x_m = \frac{\sum_{i=1}^{n} x_i}{n} \tag{1.1}$$

Here are the selected properties of the arithmetic mean:

- The sum of the values is equal to the product of the arithmetic mean and the population size.
- The arithmetic mean fulfills the following condition:

$$x_{min} < x_m < x_{max} \tag{1.2}$$

- The sum of deviations of individual values from the mean is equal to zero:

$$\sum_{i=1}^{n} (x_i - x_m) = 0 \tag{1.3}$$

- The sum of squares of deviations of each value from the mean is minimal:

$$\sum_{i=1}^{n} (x_i - x_m)^2 = \min \tag{1.4}$$

- The arithmetic mean is sensitive to extreme values (outliers) of the characteristic.
- The arithmetic mean from a sample is a good approximation (estimation, estimator) of the expected value.

The *truncated mean* x_{wk} is a statistical measurement calculated for the series of results, among which the extrema (minima or maxima) have a high uncertainty concerning their actual value [2]. Its value is calculated according to the following formula

$$x_{wk} = \frac{1}{n} \left[(k+1)x_{(k+1)} + \sum_{i=k+2}^{n-k-1} x_{(i)} + (k+1)x_{(n-k)} \right] \tag{1.5}$$

where
  x_{wk}: Truncated mean
  n: Number of results in the series
  k: Number of extreme (discarded) results

Mode *Mo* is the value that occurs most frequently in a data set. In a set of results, there may be more than one value that can be a mode, because the same maximum frequency can be attained at different values.

Quantiles q are values in an investigated population (a population presented in the form of a statistical series) that divide the population into a certain number of subsets. Quantiles are data values marking boundaries between consecutive subsets.

The 2-quantile is called the *median*, 4-quantiles are called *quartiles*, 10-quantiles are *deciles*, and 100-quantiles are *percentiles*.

A *quartile* is any of three values that divide a sorted data set into four equal parts, so that each part represents one-quarter of the sampled population.

The first quartile (designated q_1) divides the population in such a way that 25 percent of the population units have values lower than or equal to the first quartile q_1, and 75 percent of the units have values higher than or equal to the first quartile. The second quartile q_2 is the median. The third quartile (designated q_3) divides the population in such a way that 75 percent of the population units have values lower than or equal to the third quartile q_3, and 25 percent units have values higher than or equal to the quartile.

The *median Me* measurement is the middle number in a population arranged in a nondecreasing order (for a population with an odd number of observations), or the mean of the two middle values (for those with an even number of observations).

A median separates the higher half of a population from the lower half; half of the units have values smaller than or equal to the median, and half of them have values higher than or equal to the median. Contrary to the arithmetic mean, the median is not sensitive to outliers in a population. This is usually perceived as its advantage, but sometimes may also be regarded as a flaw; even immense differences between outliers and the arithmetic mean do not affect its value.

Hence, other means have been proposed; for example, the truncated mean. This mean, less sensitive to outliers than the standard mean (only a large number of outliers can significantly influence the truncated mean) and standard deviation, is calculated using all results, which transfers the extreme to an accepted deviation range—thanks to the application of appropriate iterative procedures.

The first *decile* represents 10 percent of the results that have values lower than or equal to the first decile, and 90 percent of the results have values greater than or equal to it.

1.4 MEASURES OF DISPERSION

Measures of dispersion (variability) are usually used to determine differences between individual observations and mean value [1].

The most popular measures of dispersion are

- Range
- Variance
- Standard deviation
- Average deviation
- Coefficient of variation

The *range R* is a difference between the maximum and minimum value of an examined characteristic:

$$R = x_{max} - x_{min} \tag{1.6}$$

It is a measure characterizing the empirical variability region of the examined characteristic, but does not give information on the variability of individual values of the characteristic in the population.

Variance V is an arithmetic mean of the squared distance of values from the arithmetic mean of the population. Its value is calculated according to the formula

$$V = \frac{1}{n-1} \sum_{i-1}^{n} (x_i - x_m)^2 \tag{1.7}$$

Standard deviation SD, the square root of the variance, is the measure of dispersion of individual results around the mean. It is described by the following equation:

$$SD = \sqrt{\frac{\sum_{i=1}^{n}(x_i - x_m)^2}{n-1}} \quad (1.8)$$

SD equals zero only when all results are identical. In all other cases it has positive values. Thus, the greater the dispersion of results, the greater the value of the *SD*.

It must be remembered that dispersion of results occurs in each analytical process. Yet it is not always observed, for example, because the resolution of a measuring instrument being too low.

Properties of *SD* include the following

- If a constant value is added to or subtracted from each value, the *SD* does not change.
- If each measurement value is multiplied or divided by any constant value, the *SD* is also multiplied/divided by that same constant.
- *SD* is always a denominate number, and it is always expressed in the same units as the results.

If an expected value μ_x is known, the *SD* is calculated according to the following formula:

$$SD = \sqrt{\frac{\sum_{i=1}^{n}(x_i - \mu_x)^2}{n}} \quad (1.9)$$

Relative standard deviation (RSD) is obtained by dividing the *SD* by the arithmetic mean:

$$RSD = \frac{SD}{x_m} \quad (1.10)$$

Obviously, $x_m \neq 0$.

The *SD* of the arithmetic mean $\overline{SD}$ is calculated according to the following equation:

$$\overline{SD} = \frac{SD}{\sqrt{n}} \quad (1.11)$$

The *SD* of an analytical method SD_g (general) is determined using the results from a series of measurements:

$$SD_g = \sqrt{\frac{1}{n-k} \sum_{i=1}^{k} SD_i^2 (n_i - 1)} \qquad (1.12)$$

where k equals the number of series of parallel determinations.

For series with equal numbers of elements, the formula is simplified to the following equation:

$$SD_g = \sqrt{\frac{1}{k} \sum_{i=1}^{k} SD_i^2} \qquad (1.13)$$

The *mean absolute deviation D* is an arithmetic mean of absolute deviations of the values from the arithmetic mean. It determines the mean difference between the results in the population and the arithmetic mean:

$$D = \frac{1}{n} \sum_{i=1}^{n} |x_i - x_m| \qquad (1.14)$$

The relationship between the mean and *SDs* for the same set of results can be presented as $D < SD$.

The coefficient of variation *CV* is *RSD* presented in percentage points:

$$CV = RSD[\%] \qquad (1.15)$$

The *CV* is the quotient of the absolute variation measure of the investigated characteristic and the mean value of that characteristic. It is an absolute number, usually presented in percentage points.

The *CV* is usually applied in comparing differences:

- Among several populations with regard to the same characteristic
- Within the same population with regard to a few different characteristics

1.5 MEASURES OF ASYMMETRY

A *skewness (asymmetry) coefficient* is an absolute value expressed as the difference between an arithmetic mean and a mode.

The skewness coefficients are applied in comparisons in order to estimate the force and the direction of asymmetry. These are absolute numbers: The greater the asymmetry, the greater their value.

The quartile skewness coefficient shows the direction and force of result asymmetry located between the first and third quartiles.

1.6 MEASURES OF CONCENTRATION

A *concentration coefficient K* is a measure of the concentration of individual observations around the mean. The greater the value of the coefficient, the more slender the frequency curve and the greater the concentration of the values about the mean.

Example 1.1

Problem: For the given series of measurement results, give the following values:

- Mean
- Standard deviation
- Relative standard deviation
- Mean absolute deviation
- Coefficient of variation
- Minimum
- Maximum
- Range
- Median
- Mode

Data: Result series, mg/dm³:

1	12.34
2	12.67
3	12.91
4	12.02
5	12.52
6	12.12
7	12.98
8	12.34
9	12.00
10	12.67
11	12.53
12	12.34
13	12.79

Solution:

Mean, x_m, mg/dm³	12.48
Standard deviation, SD, mg/dm³	0.32
Relative standard deviation, RSD	0.0257
Mean absolute deviation, D	0.264
Coefficient of variation—$CV,\%$	2.57%
Minimum, x_{min}, mg/dm³	12.00
Maximum, x_{max}, mg/dm³	12.98
Range, R, mg/dm³	0.98
Median, Me, mg/dm³	12.52
Mode, Mo, mg/dm³	12.34

Excel file: exampl_stat01.xls

1.7 STATISTICAL HYPOTHESIS TESTING

A *hypothesis* is a proposition concerning a population, based on probability, assumed in order to explain some phenomenon, law, or fact. A hypothesis requires testing.

Statistical hypothesis testing means checking propositions with regard to a population that have been formulated without examining the whole population. The plot of the testing procedure involves:

1. Formulating the null hypothesis and the alternative hypothesis. The null hypothesis H_o is a simple form of the hypothesis that is subjected to tests. The alternative hypothesis H_1 is contrasted with the null hypothesis.
2. The choice of an appropriate test. The test serves to verify the hypothesis.
3. Determination of the level of significance α.
4. Determining the critical region of a test. The size of the critical region is determined by any low level of significance α, and its location is determined by the alternative hypothesis.
5. Calculation of a test's parameter using a sample. The results of the sample are processed in a manner appropriate to the procedure of the selected test and are the basis for the calculation of the test statistic.
6. Conclusion. The test statistic, determined using the sample, is compared with the critical value of the test:
   - If the value falls within the critical region, then the null hypothesis should be rejected as false. It means that the value of the calculated test parameter is greater than the critical value of the test (read from a relevant table).
   - If the value is outside the critical region, it means that there is not enough evidence to reject the null hypothesis. It means that the value of the calculated parameter is not greater than the critical value of the test (read from a relevant table); hence, the conclusion that the null hypothesis may be true.

   Errors made during verification:
   - Type I error: Incorrectly rejecting the null hypothesis H_o when it is true
   - Type II error: Accepting the null hypothesis H_o when it is false

Nowadays, statistical hypothesis testing is usually carried out using various pieces of software (e.g., Statistica®). In this case, the procedure is limited to calculating the parameter p for a given set of data after selecting an appropriate statistical test. The value p is then compared with the assumed value of the level of significance α.

If the calculated value p is smaller than the α value ($p < \alpha$), the null hypothesis H_o is rejected. Otherwise, the null hypothesis is not rejected.

The basic classification of a statistical test divides tests into parametric and nonparametric ones.

Parametric tests serve to verify parametric hypotheses on the distribution parameters of the examined characteristic in a parent population. Usually they are used to test propositions concerning arithmetic mean and variance. The tests are constructed with the assumption that the CDF is known for the parent population.

Nonparametric tests are used to test various hypotheses on the goodness of fit in one population with a given theoretical distribution, the goodness of fit in two populations, and the randomness of sampling.

1.8 STATISTICAL TESTS

During the processing of analytical results, various statistical tests can be used. Their descriptions, applications, and inferences based on these tests are presented below. Appropriate tables with critical values for individual tests are given in the appendix at the end of the book.

1.8.1 Confidence Interval Method [3]

Test	Confidence Interval Method
Aim	Test whether a given set of results includes a result(s) with a gross error
Requirements	• Set size 3–10 • Unbiased series—an initially rejected uncertain result • Only one result can be rejected from a given set
Course of action	• Exclude from a set of results the result that was initially recognized as one with a gross error • Calculate the endpoints of the confidence interval for a single result based on the following formula:

$$g = x_m \pm t_{crit} \sqrt{\frac{n}{n-2}} SD \qquad (1.16)$$

where
x_m: Mean for an unbiased series
SD: Standard deviation for an unbiased series
n: Entire size of a series, together with an uncertain result
t_{crit}: Critical parameter of the Student's t test, read for $f = n - 2$ degrees of freedom—Table A.1 (in the appendix)

Inference	If an uncertain result falls outside the limits of the confidence interval, it is rejected; otherwise, it is compensated for in further calculations and the values of x_m and SD are calculated again
Requirements	• Set size is 3–10 • Unbiased series—an initially rejected doubtful result • Only one result can be rejected from a given set

Course of action	• Exclude from a set of results the result that was initially recognized as one with a gross error • Calculate the value of the parameter t_{calc} according to the following formula:

$$t_{calc} = \frac{|x_i - x_m|}{SD} \tag{1.17}$$

where
x_i: Uncertain result
x_m: Mean value for the unbiased series
SD: Standard deviation for the unbiased series

• Compare the value of t_{calc} with the critical value calculated according to the following formula

$$t_{crit(corr)} = t_{crit} \cdot \sqrt{\frac{n}{n-2}} \tag{1.18}$$

where
n: Entire size of a series, together with an uncertain result
t_{crit}: Critical parameter of the Student's t test, read for $f = n - 2$ degrees of freedom—Table A.1 (in the appendix)

Inference	If $t_{calc} \leq t_{crit(corr)}$, then the initially rejected result is included in further calculations and x_m and s are calculated again; otherwise the initially rejected result is considered to have a gross error
Requirements	• Set size 3–10 • Unbiased series—an initially rejected uncertain result • Only one result can be rejected from a given set
Course of action	Calculate the endpoints of the confidence interval for an individual result using the following formula

$$g = x_m \pm w_\alpha \cdot SD \tag{1.19}$$

where
x_m: Mean for the unbiased series
SD: Standard deviation for the unbiased series
w_α: Critical parameter determined for the number of degrees of freedom $f = n - 2$: Table A.2 (in the appendix)
n: Total number of a series

Inference	If the uncertain result falls outside the endpoints of the determined confidence interval, it is rejected and x_m and SD are calculated again

Requirements
- Set size >10
- Biased series

Course of action
- Calculate the endpoints of the confidence interval for an individual result using the following formula:

$$g = x_m \pm k_\alpha \cdot SD \tag{1.20}$$

where

x_m: Mean for the biased series

SD: Standard deviation for the biased series

k_α: Confidence coefficient for a given level of significance α, from a normal distribution table:

for $\alpha = 0.05$ $k_\alpha = 1.65$
for $\alpha = 0.01$ $k_\alpha = 2.33$

Inference   If the uncertain result(s) falls outside the endpoints of the determined confidence interval, it is rejected and x_m and SD are calculated again

Requirements
- Set size >10
- Unbiased series: An initially rejected uncertain result
- Known value of the method's standard deviation

Course of action
- Calculate the endpoints of the confidence interval for an individual result using the following formula:

$$g = x_m \pm k_\alpha \cdot SD_g \sqrt{\frac{n}{n-1}} \tag{1.21}$$

where

x_m: Mean for the unbiased series

SD_g: Standard deviation of the method

k_α: Confidence coefficient for a given level of significance α, from a normal distribution table:

for $\alpha = 0.05$ $k_\alpha = 1.65$
for $\alpha = 0.01$ $k_\alpha = 2.33$

Inference   If the uncertain result falls outside the endpoints of the determined confidence interval, it is rejected; otherwise, it is included in the series and x_m and SD are calculated again

1.8.2 CRITICAL RANGE METHOD [3]

Test	**Critical Range Method**
Aim	Test whether a given set of results includes a result(s) with a gross error
Requirements	• Set size >10
	• Known value of the method's standard deviation: SD_g
Course of action	• Calculate the value of the range result according to the following formula:

$$R = x_{max} - x_{min}$$

• Calculate the value of the critical range according to the following formula:

$$R_{crit} = z \cdot SD_g \tag{1.22}$$

where
SD_g: The standard deviation of the method
z: Coefficient from the table for a given level of confidence α and n parallel measurements and f degrees of freedom: Table A.3 (see the appendix)

Inference	If $R > R_{crit}$, the extremum result is rejected and the procedure is conducted anew
Requirements	• Known value of the mean range for the series—R_m
	• Known results of k series of parallel determinations, with n determinations in each series (most often $n = 2$ or 3; $k \geq 30$)
Course of action	• Calculate the value of the range for each series according to the following formula:

$$R_i = x_{max_i} - x_{min_i} \tag{1.23}$$

• Calculate the value of the critical range according to the following formula:

$$R_{crit} = z_\alpha \cdot R_m \tag{1.24}$$

where
z_α: Coefficient from a table for a given level of confidence α and n parallel measurements in a series: Table A.4 (in the appendix)

Inference	If $R_i > R_{crit}$, the i-th series of the measurement results is rejected

1.8.3 Dixon's Q Test [3,4]

Test	Dixon's Q Test
Aim	Test whether a given set of results includes a result with a gross error
Hypotheses	H_o: In the set of results there is no result with a gross error
	H_1: In the set of results there is a result with a gross error
Requirements	• Set size 3–10
	• Test whether a given set of results includes a result with a gross error
Course of action	• Order the results in a non-decreasing sequence: $x_1...x_n$
	• Calculate the value of the range R according to the formula $R = x_n - x_1$
	• Calculate the value of parameters Q_1 and Q_n according to the formulas

$$Q_1 = \frac{x_2 - x_1}{R} \quad Q_n = \frac{x_n - x_{n-1}}{R} \tag{1.25}$$

	• Compare the obtained values with the critical value Q_{crit} for the selected level of significance α and the number of degrees of freedom $f = n$, Table A.5 (in appendix)
Inference	If one of the calculated parameters exceeds the critical value Q_{crit}, then the result from which it was calculated (x_n or x_1) should be rejected as a result with a gross error and only then should x_m and SD be calculated

In some studies [1], the authors use a certain type of Dixon's Q test that makes it possible to test a series comprising up to 40 results.

Test	Dixon's Q Test
Aim	Test whether a given set of results includes a result with a gross error
Hypotheses	H_o: In the set of results there is no result with a gross error
	H_1: In the set of results there is a result with a gross error
Requirements	• Set size 3–7
	• Test whether a given set of results includes a result with a gross error

Course of action
- Order the results as a non-decreasing sequence: $x_1...x_n$
- Calculate the value of the range R according to the formula $R = x_n - x_1$
- Calculate the value of parameters Q_1 and Q_n according to the formulas

$$Q_1 = \frac{x_2 - x_1}{R} \quad Q_n = \frac{x_n - x_{n-1}}{R} \tag{1.26}$$

- Compare the obtained values with the critical value Q_{crit} for the selected level of significance α and the number of degrees of freedom $f = n$, Table A.6 (in appendix)

Requirements
- Set size 8–12
- Test whether a given set of results includes a result with a gross error

Course of action
- Order the results as a non-decreasing sequence: $x_1...x_n$
- Calculate the value of parameters Q_1 and Q_n according to the formulas

$$Q_1 = \frac{x_2 - x_1}{x_{n-1} - x_1} \quad Q_n = \frac{x_n - x_{n-1}}{x_n - x_2} \tag{1.27}$$

- Compare the obtained values with the critical value Q_{crit} for the selected level of significance α and the number of degrees of freedom $f = n$, Table A.6 (in appendix)

Requirements
- Set size > 12
- Test whether a given set of results includes a result with a gross error

Course of action
- Order the results as a non-decreasing sequence: $x_1...x_n$
- Calculate the value of parameters Q_1 and Q_n according to the formulas

$$Q_1 = \frac{x_2 - x_1}{x_{n-1} - x_1} \quad Q_n = \frac{x_n - x_{n-1}}{x_n - x_2} \tag{1.28}$$

- Compare the obtained values with the critical value Q_{crit} for the selected level of significance α and the number of degrees of freedom $f = n$, Table A.6 (in appendix)

Inference
If one of the calculated parameters exceeds the critical value Q_{crit}, then the result from which it was calculated (x_n or x_1) should be rejected as a result with a gross error and only then should x_m and SD be calculated

1.8.4 CHI SQUARE TEST [3]

Test	**Chi Square (χ^2) Test**
Aim	Test if the variance for a given series of results is different from the set value
Hypotheses	H_o: The variance calculated for the series of results is not different from the set value in a statistically significant manner
	H_1: The variance calculated for the series of results is different from the set value in a statistically significant manner
Requirements	Normal distribution of results in a series
Course of action	• Calculate the standard deviation for the series of results
	• Calculate the chi square test parameter χ^2 according to the formula

$$\chi^2 = \frac{n \cdot SD^2}{SD_o^2} \qquad (1.29)$$

where

SD: The standard deviation calculated for the set of results

SD_o: The set value of the standard deviation

n: The number of results in an investigated set

• Compare the calculated value χ^2 with the critical value χ^2_{crit} for the assumed level of significance α and the calculated number of degrees of freedom $f = n - 1$; Table A.7 (in the appendix)

Inference

• If the calculated value χ^i does not exceed the critical value $\left(\chi^2 \leq \chi^2_{crit}\right)$, then it may be inferred that the calculated value of the standard deviation does not differ in a statistically significant manner from the set value—acceptance of hypothesis H_o

• If the calculated value χ^2 is greater than the critical value read from the tables $\left(\chi^2 > \chi^2_{crit}\right)$, then it may be inferred that the compared values of the standard deviation differ in a statistically significant manner—rejection of the hypothesis H_o

1.8.5 SNEDECOR'S *F* TEST [3–5]

Test	**Snedecor's *F* Test**
Aim	Compare the standard deviations (variances) for two sets of results

Hypotheses H_o: The variances calculated for the compared series of results do not differ in a statistically significant manner

H_1: The variances calculated for the compared series of results differ in a statistically significant manner

Requirements   Normal distributions of results in a series

Course of action   • Calculate the standard deviations for the compared series of results

• Calculate Snedecor's F test parameter according to the formula

$$F = \frac{SD_1^2}{SD_2^2} \qquad (1.30)$$

where

SD_1, SD_2: Standard deviations for the two sets of results

Note: The value of the expression should be constructed in such a way so that the numerator is greater than the denominator: The value F should always be greater than 1

• Compare the calculated value with the critical value of the with an assumed level of significance α and the calculated number of freedom degrees f_1 and f_2 (where $f_1 = n_1 - 1$ and $f_2 = n_2 - 1$)—Table A.8 (in the appendix)

Inference   • If the calculated value F does not exceed the critical value ($F \le F_{crit}$), then it may be inferred that the calculated values for the standard deviation do not differ in a statistically significant manner—acceptance of the hypothesis H_o

• If the calculated value F is greater than the critical value read from the tables ($F > F_{crit}$), then it may be inferred that the compared values for the standard deviation differ in a statistically significant manner—rejection of the hypothesis H_o

1.8.6   HARTLEY'S F_{MAX} TEST [3]

Test   **Hartley's F_{max} Test**

Aim   Compare the standard deviations (variances) for many sets of results

Hypotheses H_o: The variances calculated for the compared series of results do not differ in a statistically significant manner

H_1: The variances calculated for the compared series of results differ in a statistically significant manner

Requirements
- Normal distributions of results in a series
- Numbers of results in each series of the sets greater than 2
- Set sizes are identical
- The number of series not greater than 11

Course of action
- Calculate the standard deviations for the compared series of results
- Calculate the value of the F_{max} test parameter according to the following formula:

$$F_{max} = \frac{SD^2_{max}}{SD^2_{min}} \tag{1.31a}$$

where

SD_{max}, SD_{min}: The greatest and smallest value from the calculated standard deviations for the sets of results

In the case of different values of results in the series use CV instead of SD according to the following formula:

$$F_{max} = \frac{CV^2_{max}}{CV^2_{min}} \tag{1.31b}$$

- Compare the calculated value with the critical value of the parameter for the assumed level of significance α, the calculated number of degrees of freedom $f = n - 1$, and the number of the compared series k—Table A.9 (in the appendix)

Inference
- If the calculated value F_{max} does not exceed the critical value $\left(F_{max} \leq F_{max_o} \right)$, then it may be inferred that calculated standard deviations do not differ in a statistically significant manner—acceptance of the hypothesis H_o
- If the calculated value F_{max} is greater than the critical value read from the tables $\left(F_{max} > F_{max_o} \right)$, then it may be inferred that the compared standard deviations differ in a statistically significant manner—rejection of the hypothesis H_o

1.8.7 BARTLETT'S TEST [3]

Test   **Bartlett's Test**

Aim   Compare the standard deviations (variances) for many sets of results

Hypotheses	H_o: The variances calculated for the compared series of results do not differ in a statistically significant manner H_1: The variances calculated for the compared series of results differ in a statistically significant manner
Requirements	The number of results in each series of the sets is greater than 2
Course of action	• Calculate the standard deviation for the compared series of results • Calculate the value of a Q test parameter according to the following formula

$$Q = \frac{2.303}{c}\left[(n-k)\log\left(\overline{SD_o^2}\right) - \sum_{i=1}^{k}(n_i-1)\log\left(SD_i^2\right)\right] \qquad (1.32)$$

in which

$$c = 1 + \frac{1}{3(k-1)}\left(\sum_{i=1}^{k}\frac{1}{n_i-1} - \frac{1}{n-k}\right) \qquad (1.33)$$

$$\overline{SD_o^2} = \frac{1}{n-k}\sum_{n=1}^{k}SD_i^2(n_i-1) \qquad (1.34)$$

where
n: The total number of parallel determinations
k: The number of the compared method (series)
n_i: The number of parallel determinations in a given series
SD_i: The standard deviation for the series i

• Compare the calculated value with the critical value of the χ_{crit}^2 parameter for the assumed level of significance α and the calculated number of degrees of freedom $f = k - 1$—Table A.7 (in the appendix)

Inference

• If the calculated value Q does not exceed the critical value $\left(Q \leq \chi_{crit}^2\right)$, then it may be inferred that the calculated standard deviations do not differ in a statistically significant manner—acceptance of the hypothesis H_o

• If the calculated value Q is greater than the critical value read from the tables $\left(Q > \chi_{crit}^2\right)$, then it may be inferred that the compared standard deviations differ in a statistically significant manner—rejection of the hypothesis H_o

1.8.8 MORGAN'S TEST [3]

Test   **Morgan's Test**

Aim   Compare standard deviations (variances) for two sets of
                        dependent (correlated) results

Hypotheses H_o: The variances calculated for the compared series of results
                        do not differ in a statistically significant manner
                        H_1: The variances calculated for the compared series of results
                        differ in a statistically significant manner

Requirements   Number of results in each series of the sets is greater than 2

Course of action   • Calculate the standard deviations for the compared series of
                          results
                        • Calculate the regression coefficient r according to the
                          following formula

$$r = \frac{k\sum_{i=1}^{k} x_{1i}x_{2i} - \sum_{i=1}^{k} x_{1i} \sum_{i=1}^{k} x_{2i}}{\sqrt{\left[k\sum_{i=1}^{k} x_{1i}^2 - \left(\sum_{i=1}^{k} x_{1i}\right)^2\right]\left[k\sum_{i=1}^{k} x_{2i}^2 - \left(\sum_{i=1}^{k} x_{2i}\right)^2\right]}} \qquad (1.35)$$

• Calculate the value of testing L parameter according to the
  following formula:

$$L = \frac{4SD_1^2 SD_2^2 (1-r^2)}{\left(SD_1^2 + SD_2^2\right) - 4r^2 SD_1^2 SD_2^2} \qquad (1.36)$$

• Calculate the value of parameter t according to the
  following formula:

$$t = \sqrt{\frac{(1-L)(k-2)}{L}} \qquad (1.37)$$

where
k: The number of pairs of results
x_{1i}, x_{2i}: Individual values of results for the compared sets
• Compare the calculated value t with the critical value t_{crit}, a
  parameter for the assumed level of significance α the
  calculated number of degrees of freedom $f = k - 2$—Table
  A.1 (in the appendix)

Inference
- If the calculated value t does not exceed the critical value t_{crit}, so that the relation $t \le t_{crit}$ is satisfied, then it may be inferred that the calculated standard deviations do not differ in a statistically significant manner—acceptance of hypothesis H_o
- If the calculated value t is greater than the critical value read from the tables ($t > t_{crit}$), then it may be inferred that the compared standard deviations differ in a statistically significant manner—rejection of the hypothesis H_o

1.8.9 STUDENT'S t TEST [3,4]

Test	**Student's t Test**
Aim	Compare means for two series (sets) of results
Hypotheses	H_o: The calculated means for the compared series of results do not differ in a statistically significant manner
	H_1: The calculated means for the compared series of results differ in a statistically significant manner
Requirements	• Normal distributions of results in a series
	• Number of results in each series of the sets greater than 2
	• Insignificant variance differences for the compared sets of results—Snedecor's F test, Section 1.8.5
Course of action	• Calculate the means and standard deviations for the series of results
	• Calculate Student's t test parameter according to the following equation

$$t = \frac{|x_{1m} - x_{2m}|}{\sqrt{(n_1 - 1)SD_1^2 + (n_2 - 1)SD_2^2}} \sqrt{\frac{n_1 n_2 (n_1 + n_2 - 2)}{n_1 + n_2}} \qquad (1.38)$$

where

x_{1m}, x_{2m}: The means calculated for the two compared sets of results

SD_1, SD_2: The standard deviations for the sets of results
- Compare the calculated value with the critical value of a parameter for the assumed level of significance α and the calculated number of degrees of freedom $f = n_1 + n_2 - 2$—Table A.1 (in the appendix)

Inference	• If the value *t* does not exceed the critical value t_{crit} ($t \leq t_{crit}$), then it may be inferred that the obtained means do not differ in a statistically significant manner—acceptance of the hypothesis H_o
	• If the calculated value *t* is greater than the critical value read from the tables ($t > t_{crit}$), then it is inferred that the compared means differ in a statistically significant manner—rejection of the hypothesis H_o
Test	**Student's *t* Test**
Aim	Compare the mean with the assumed value
Hypotheses	H_o: The calculated mean does not differ in a statistically significant manner from the assumed value
	H_1: The calculated mean differs in a statistically significant manner from the assumed value
Requirements	• Normal distribution of results in a series
	• The number of results in a series of sets is greater than 2
Course of action	• Calculate the mean and standard deviation for the series of results
	• Calculate Student's *t* test parameter according to the following equation:

$$t = \frac{|x_m - \mu|}{SD} \sqrt{n} \qquad (1.39)$$

where

x_m: The mean calculated for the set of results

μ: The reference (e.g., certified value)

SD: The unit of deviation, for example, the standard deviation of the set of results which the mean was calculated based on

n: The number of results

• Compare the calculated value with the critical value of a parameter, for the assumed level of significance α, the calculated number of degrees of freedom $f = n - 1$—Table A.1 (in the appendix)

Inference	• If the value *t* does not exceed the critical value t_{crit} ($t \leq t_{crit}$), then it may be inferred that the obtained mean is not different from the set value in a statistically significant manner—acceptance of the hypothesis H_o
	• If the calculated value *t* is greater than the critical value read from the tables ($t > t_{crit}$), it is inferred that the mean is different from the set value in a statistically significant manner—rejection of the hypothesis H_o

1.8.10 COCHRAN–COX C TEST [3]

Test	**Cochran–Cox C Test**
Aim	Compare the means for the series of sets of results, for which the standard deviations (variances) differ in a statistically significant manner
Hypotheses	H_o: The calculated means for the compared series of results do not differ in a statistically significant manner
	H_1: The calculated means for the compared series of results differ in a statistically significant manner
Requirements	• Normal distribution of results in a series
	• The number of results in a series of sets is greater than 2
Course of action	• Calculate the means and standard deviations for the compared series of results
	• Calculate the value of a parameter C according to the following formula:

$$C = \frac{|x_{1m} - x_{2m}|}{\sqrt{z_1 + z_2}} \tag{1.40}$$

in which

$$z_1 = \frac{SD_1^2}{n_1 - 1}, \text{ and } z_2 = \frac{SD_2^2}{n_2 - 1} \tag{1.41}$$

where

x_{1m}, x_{2m}: The means calculated for the two compared sets of results

SD_1, SD_2: The standard deviations for the sets of results

• Calculate the critical value of the parameter C (C_{crit}) according to the following formula:

$$C_{crit} = \frac{z_1 t_1 + z_2 t_2}{z_1 + z_2} \tag{1.42}$$

where

t_1 and t_2: The critical values read from the tables of Student's t distribution (Table A.1) respectively for $f_1 = n_1 - 1$ and $f_2 = n_2 - 1$, the number of degrees of freedom and the level of significance α

• Compare the calculated value C with the calculated critical value C_{crit}

Inference	• If the value C does not exceed the critical value C_{crit} ($C \le C_{crit}$), then it may be inferred that the obtained mean values do not differ from one another in a statistically significant manner—acceptance of the hypothesis H_o
	• If the calculated value C is greater than the calculated critical value ($C > C_{crit}$), then it is inferred that the obtained means differ from one another in a statistically significant manner—rejection of the hypothesis H_o

1.8.11 ASPIN–WELCH TEST [3]

Test	**Aspin–Welch Test**
Aim	Compare the means for the series of sets of results for which the standard deviations (variances) differ in a statistically significant manner
Hypotheses	H_0: Calculated means for the compared series of results do not differ in a statistically significant manner
	H_1: Calculated means for the compared series of results differ in a statistically significant manner
Requirements	• Normal distribution of results in a series
	• The number of results in a series of sets is greater than 6
Course of action	• Calculate the means and standard deviations for the compared series of results
	• Calculate the values of expressions described using the following equations:

$$v = \frac{|x_{1m} - x_{2m}|}{\sqrt{\dfrac{SD_1^2}{n_1} + \dfrac{SD_2^2}{n_2}}} \tag{1.43}$$

$$c = \frac{\dfrac{SD_1^2}{n_1}}{\dfrac{SD_1^2}{n_1} + \dfrac{SD_2^2}{n_2}} \tag{1.44}$$

in which

$$\frac{SD_1^2}{n_1} < \frac{SD_2^2}{n_2} \tag{1.45}$$

where

x_{1m}, x_{2m}: The means calculated for the two compared sets of results

SD_1, SD_2: The standard deviations for the sets of results

- Compare the calculated value v with the critical value v_o for the corresponding level of significance α, the number of degrees of freedom $f_1 = n_1 - 1, f_2 = n_2 - 1$, and the calculated values of c, and thus v_o (α, f_1, f_2, c)—Table A.10 (in the appendix)

Inference
- If the value v does not exceed the critical value v_o, $(v \le v_o)$, then it may be inferred that the obtained means do not differ from one another in a statistically significant manner—acceptance of the hypothesis H_o
- If the calculated value v is greater than the calculated critical value $(v > v_o)$, it is inferred that the obtained means differ from one another in a statistically significant manner—rejection of the hypothesis H_o

1.8.12   Cochran's Test [6]

Test   **Cochran's Test**

Aim   Detection of outliers in a given set—intralaboratory variability test
One-sided test for outliers—the criterion of the test examines only the greatest standard deviations

Requirements
- The number of results in a series (set) greater than or equal to 2, but only when the number of compared laboratories is greater than 2
- Sets of results (series) with the same numbers
- It is recommended to apply the tests before the Grubbs' test—Section 1.8.13

Course of action
- Calculate the standard deviations for each of the compared sets of results
- Calculate the value of parameter C using the following formula:

$$C = \frac{SD_{max}^2}{\sum_{i=1}^{p} SD_i^2} \tag{1.46}$$

where

SD_{max}: Maximum standard deviation in the investigated set (among the investigated laboratories)

SD_i: The standard deviation for a given series (data from a laboratory)

p: The number of standard deviations (the number of compared laboratories)

- Compare the calculated value C with the critical value for a given value n, the number of results in a series and p, the number of laboratories—Table A.11 (in the appendix)

Inference
- If the value of the calculated test parameter is less than or equal to the critical value corresponding to the level of significance $\alpha = 0.05$, then the investigated result is considered to be correct
- If the numerical value of a respective test parameter is greater than the critical value corresponding to the level of significance $\alpha = 0.05$ and less than or equal to the critical value corresponding to the level of significance $\alpha = 0.01$, then the investigated result is an uncertain value
- If the value of the test parameter is greater than the critical value corresponding to the level of significance $\alpha = 0.01$, then the investigated result is considered to be an outlier

1.8.13 GRUBBS' TEST [6,7]

Test	**Grubbs' Test**
Aim	Detect outliers in a given set—interlaboratory variability test
Requirements	

- The number of results in a series (set) is greater than or equal to 2, but only when the number of compared laboratories is greater than 2
- The same number of results in the sets (series) of results
- It is recommended to apply this test before Cochran's test—Section 1.8.12
- With a single use it enables the detection of one outlier; thus, it should be repeated until no outliers are observed in the remaining results

Course of action
- Calculate the standard deviation for the set of results
- Order the set of data x_i for $I = 1, 2,..., p$ in an increasing sequence
- Calculate the value of parameter G_p according to the following relation:

$$G_p = \frac{(x_p - x_m)}{SD} \tag{1.47}$$

where
x_p: The value in the set of results considered to be an outlier
x_m: The mean
SD: The standard deviation
- Compare the calculated value G_p with the critical value for a given value p, the number of laboratories—Table A.12 (in the appendix)

Inference

- If the value of the calculated test parameter is less than or equal to the critical value corresponding to the level of significance $\alpha = 0.05$, then the investigated result is considered to be correct
- If the numerical value of a corresponding test parameter is greater than the critical value corresponding to the level of significance $\alpha = 0.05$, and less than or equal to the critical value corresponding to the level of significance $\alpha = 0.01$, then the investigated result is an uncertain value
- If the value of the test parameter is greater than the critical value corresponding to the level of significance $\alpha = 0.01$, then the investigated result is considered to be an outlier; after rejection of this value from the set of results, the test for the series of $p - 1$ results may be conducted again, and the course of action should be continued until there are no more outliers in the set of results

Test

Grubbs' Test

Aim

Detect outliers in a given set—interlaboratory variability test

Requirements

- The number of results in a series (set) is greater than or equal to 2, but only when the number of compared laboratories is greater than 2
- The same number of results in the sets of results (series)
- It is recommended to apply this test before Cochran's test—Section 1.8.12
- In a given course of action two (the greatest or the smallest) results may be rejected from the set of results

Course of action

- Order the set of data x_i for $I = 1, 2,..., p$ in an increasing sequence
- Calculate the values of parameter SD_o according to the following equation:

$$SD_o^2 = \sum_{i=1}^{p} (x_i - x_m)^2 \tag{1.48}$$

- Calculate the values of parameters $x_{m(p-1,p)}$ when testing two of the highest results or $x_{m(1,2)}$, two of the lowest results, according to the following equations:

$$x_{m(p-1,p)} = \frac{1}{p-1}\sum_{i=1}^{p-2} x_i, \text{ or } x_{m(1,2)} = \frac{1}{p-1}\sum_{i=3}^{p} x_i \tag{1.49}$$

- Calculate the values of respective parameters: $SD_{(p-1,p)}$ or $SD_{(1,2)}$ according to the following equations:

$$SD^2_{(p-1,p)} = \sum_{i=1}^{p-2}\left(x_i - x_{m(p-1,p)}\right), \text{ or } SD^2_{(2,1)} = \sum_{i=3}^{p}\left(x_i - x_{m(1,2)}\right)^2$$

$$(1.50)$$

- Calculate the value of parameter G according to the following equations:

$$G = \frac{SD^2_{(p-1,p)}}{SD^2_o}, \text{ or } G = \frac{SD^2_{(1,2)}}{SD^2_o}$$

$$(1.51)$$

- Compare the calculated value of G with the critical value for a given value p, the number of laboratories—Table A.12 (in the appendix)

Inference
- If the value of the calculated test parameter is less than or equal to the critical value corresponding to the level of significance $\alpha = 0.05$, then the investigated results are considered to be correct
- If the numerical value of a corresponding test parameter is greater than the critical value corresponding to the level of significance $\alpha = 0.05$ and less than or equal to the critical value corresponding to the level of significance $\alpha = 0.01$, then the investigated results are uncertain
- If the value of a test parameter is greater than the critical value corresponding to the level of significance $\alpha = 0.01$, then the investigated results are considered to be outliers; after rejection of these values from the set of results, the test for the series of $p - 2$ results may conducted again, and the course of action should be continued until there are no more outliers in the set of results

1.8.14 HAMPEL'S TEST

Hampel's test is called *Huber's test* by some authors [8,9].

Test	**Hampel's Test**
Aim	Detect outliers in a given set
Requirements	The number of results in a series (set) is greater than 2
Course of action	• Order the values in an increasing sequence
	• Calculate the median *Me* for all the results x_i, where x_i includes the interval from x_1 to x_n
	• Calculate the deviations of r_i from the median for each result using the following formula

$$r_i = (x_i - Me) \tag{1.52}$$

- Calculate the absolute values $|r_i|$
- Order the values of $|r_i|$ in an increasing sequence
- Calculate the median deviations $Me_{|r_i|}$
- Compare the values of $|r_i|$ with $4{,}5 \cdot Me_{|r_i|}$

Inference   If the following condition is satisfied

$$|r_i| \geq 4{,}5 \cdot Me_{|r_i|} \tag{1.53}$$

then the result x_i is considered to be an outlier

1.8.15   Z-SCORE [10,11]

Test   **Z-Score**

Aim   Detect uncertain results and outliers

Applied during the processing of results of interlaboratory comparisons

Requirements   The number of results in a series (set) greater than 2

Course of action   • Calculate the Z-score using the following formula:

$$Z = \frac{x_{lab} - x_{ref}}{SD} \tag{1.54}$$

where

x_{lab}: Result obtained by a given laboratory

x_{ref}: The assumed value/the reference value

SD: The deviation unit

The standard deviation calculated using all the values in a set; the modified standard deviation calculated using the following relation:

$$SD_{mod} = \sqrt{SD^2 + u_{(x_{ref})}^2} \tag{1.55}$$

where

$u_{(x_{ref})}$: Standard uncertainty of the accepted value/reference value

Combined standard uncertainty is calculated using the following relation:

$$u = \sqrt{u_{(x_{lab})}^2 + u_{(x_{ref})}^2} \tag{1.56}$$

where

$u_{(x_{lab})}$: Standard uncertainty of a value obtained by a given laboratory

Inference
- If $|Z| \le 2$, a result is considered to be satisfactory
- If $2 < |Z| < 3$, a result is considered to be uncertain
- If $|Z| \ge 3$, a result is considered to be unsatisfactory

1.8.16 E_n SCORE [10,11]

Test   **E_n Score**

Aim   Estimation of results of interlaboratory comparisons

Requirements   The number of results in a series (set) is greater than 2

Course of action   • Calculate the E_n score using the following formula:

$$E_n = \frac{x_{lab} - x_{ref}}{\sqrt{U^2_{(x_{lab})} + U^2_{(x_{ref})}}} \qquad (1.57)$$

where
x_{lab}: The value obtained by a given laboratory
x_{ref}: The reference value
$u(x_{lab})$: The expanded standard uncertainty result obtained by a given laboratory
$u(x_{ref})$: The expanded standard uncertainty of the reference values

Inference
- If $|E_n| \le 1$, the estimation is satisfactory
- If $|E_n| > 1$, the estimation is unsatisfactory

1.8.17 MANDEL'S h TEST [6,12,13]

Test   **Mandel's h Test**

Aim   Determine the interlaboratory traceability of results

Requirements   The number of results in a series (set) is greater than 2

Course of action   • Calculate the means x_{mi} for each series of results for each laboratory
                              • Calculate the mean for results from a given series according to the following formula:

$$\overline{x_m} = \frac{\sum_{i=1}^{p} n_i \cdot x_{mi}}{\sum_{i=1}^{p} n_i} \qquad (1.58)$$

where

n_i: The number of results for a given series obtained by a given laboratory

p: The number of laboratories

- Calculate the values of parameter h_i for a given series and for a given laboratory, according to the following formula:

$$h_i = \frac{x_{mi} - \overline{x_m}}{\frac{1}{(p-1)}\sum_{i=1}^{p}(x_{mi} - \overline{x_m})^2} \qquad (1.59)$$

- Make a graph of the values of parameter h_i for each series in the sequence of laboratories
- On the graph of the values of parameter h, mark the horizontal lines to test the data's configuration, corresponding to the Mandel h coefficients for a given level of significance ($\alpha = 0.01$ or 0.05)—Table A.13a or Table A.13b (in the appendix)

Inference   The value of parameter h_i greater than h value need to be checked from analytical viewpoint

Test	**Mandel's k Test**
Aim	Determine the interlaboratory traceability of results
Requirements	The number of results in a series (set) is greater than 2
Course of action	• Calculate the standard deviations SD_i for each series of results for each laboratory

- Calculate the values of parameter k_i for a given series and for a given laboratory, according to the following formula:

$$k_i = \frac{SD_i\sqrt{p}}{\sqrt{\sum SD_i^2}} \qquad (1.60)$$

- Make a graph of the values of parameter k_i for each series in the sequence of laboratories
- On the graph of the values of parameter k, mark the horizontal lines to test the data's configuration, corresponding to the Mandel k coefficients for a given level of significance ($\alpha = 0.01$ or 0.05)—Table A.13a or Table A.13b (in the appendix)

Inference   The value of parameter k_i greater than k value need to be checked from analytical viewpoint

1.8.18 KOLMOGOROV–SMIRNOV TEST [2,14]

Test	**Kolmogorov–Smirnov Test**
Aim	Compare the distribution of two series of results
Requirements	Two series of results
Course of action	• Calculate the empirical distribution functions for each series of results
	• Calculate the values of parameter λ_n, according to the following formula:

$$\lambda_n = \sqrt{\frac{n_1 n_2}{n_1 + n_2}} D_{n_1 n_2} \qquad (1.61)$$

where

n_1, n_2: The number of results for a given series

$D_{n_1 n_2}$: The greatest value of differences between empirical distribution functions

• Compare the λ_n value with critical value λ_α for a given level of significance—Table A.14 (in the appendix)

Inference • If the value λ_n does not exceed the critical value λ_α, ($\lambda_n \leq \lambda_\alpha$), then it may be inferred that there are no statistically significant differences in distribution functions for both compared series

• If the value λ_n does exceed the critical value λ_α, ($\lambda_n > \lambda_\alpha$), then it may be inferred that there are statistically significant differences in distribution functions for both compared series

1.9 LINEAR REGRESSION

Linear correlation is the most frequent correlation used in analytical chemistry. A decisive majority of analytical measurements uses the calibration stage, in which the values of the output signal are assigned to corresponding values of analyte concentration. To determine the functional dependency that connects the output signal with analyte concentration, a linear regression method is applied. It is also applied in determining some of the validation parameters of the analytical procedure, such as

• Accuracy: Through the determination of systematic errors
• Linearity
• Limits of detection

Therefore, we present the course of action for the linear regression method, together with a presentation of the determination method for the calibration chart parameters.

The equation of the linear regression is

$$y = b \cdot x + a \tag{1.62}$$

where
  y: Dependent variable (output signal of the measuring instrument)
  x: Independent variable (concentration of the determined analyte)
  a: Intercept
  b: Slope

The following regression parameters are calculated [3]:

- Slope:

$$b = \frac{\sum\limits_{i=1}^{n} x_i \sum\limits_{i=1}^{n} y_i - n \sum\limits_{i=1}^{n} x_i y_i}{\left(\sum\limits_{i=1}^{n} x_i\right)^2 - n \sum\limits_{i=1}^{n} x_i^2} \tag{1.63}$$

- Intercept value:

$$a = \frac{\sum\limits_{i=1}^{n} y_i - b \sum\limits_{i=1}^{n} x_i}{n} \tag{1.64}$$

- Regression coefficient:

$$r = \frac{n \sum\limits_{i=1}^{n} x_i y_i - \sum\limits_{i=1}^{n} x_i \sum\limits_{i=1}^{n} y_i}{\sqrt{\left[n \sum\limits_{i=1}^{n} x_i^2 - \left(\sum\limits_{i=1}^{n} x_i\right)^2\right]\left[n \sum\limits_{i=1}^{n} y_i^2 - \left(\sum\limits_{i=1}^{n} y_i\right)^2\right]}} \tag{1.65}$$

Values of standard deviations for

- Slope:

$$SD_b = \frac{SD_{xy}}{\sum\limits_{i=1}^{n} x_i^2 - \frac{1}{n}\left(\sum\limits_{i=1}^{n} x_i\right)^2} \tag{1.66}$$

- Intercept:

$$SD_a = SD_{xy} \sqrt{\frac{\sum\limits_{i=1}^{n} x_i^2}{n \sum\limits_{i=1}^{n} x_i^2 - \left(\sum\limits_{i=1}^{n} x_i\right)^2}} \qquad (1.67)$$

- Residuals:

$$SD_{xy} = \sqrt{\frac{\sum\limits_{i=1}^{n} (y_i - Y_i)^2}{n-2}} \qquad (1.68)$$

where
    n: The number of independent determination results for the standard solution
        samples from which the calibration curve has been determined
    y_i: The value determined experimentally
    Y_i: The value calculated from the determined regression equation

1.10 SIGNIFICANT DIGITS: RULES OF ROUNDING

A problem with correct notation of the measurement results is usually associated with issues related to significant digits and the rules of rounding.

Significant digits in the decimal notation of a given number are all the digits without initial zeros. In order to determine how many significant digits there are in a number, the number should be "read" from left to right until reaching the first digit that is not zero. That digit and all the subsequent digits are called *significant*. In the example below, the significant digits are bold:

230.546
0.00**10823**
20.1200
507.80
0.**63**×10⁴
34.70

Calculations very often use values with different numbers of significant digits and different numbers of digits after the decimal point. A value obtained from a calculation(s) should be recorded in an appropriate way, strictly dependent on the notation of the values applied in the calculation(s).

After addition or subtraction, the value of a result should be presented with the same number of digits after the decimal point as the value with the fewest number of digits after the decimal point.

For example, if a result is the sum of numbers:

$$11.23$$
$$15.2113$$
$$0.123$$
$$349.2$$

then it should be presented with one digit after the decimal point:

$$375.8$$

For multiplication and division, the number of significant digits in a result should be the same as in the value with the fewest significant digits.

If a result is a product of the following numbers:

$$11.23$$
$$15.2113$$
$$0.123$$
$$349.2$$

then it should be presented with three significant digits:

$$73.4 \times 4 \, 10^2$$

It must be remembered that the number of significant digits given in the value of a result is strictly dependent on the calculated uncertainty value (see Chapter 5). The notation of the determination requires presentation of the uncertainty value with a maximum of two significant digits and a result with the same precision (same number of figures after the decimal point) as the uncertainty value. This requirement frequently makes it necessary to round the obtained values down to the appropriate number of digits.

REFERENCES

1. Dobecki M. (ed), Zapewnienie jakości analiz chemicznych, Instytut Medycyny Pracy im. Prof. J. Nofera, Łódź, 2004 (in Polish).
2. Koronacki J., and Mielniczuk J., Statystyka dla studentów kierunków technicznych i przyrodniczych, Warsaw, WNT, 2001 (in Polish).
3. Bożyk Z., and Rudzki W., Metody statystyczne w badaniu jakości produktów żywnościowych i chemicznych, Warsaw, WNT, 1977 (in Polish).
4. Kozłowski E., Statystyczne kryteria oceny wyników i metod analitycznych w: Bobrański B.: Analiza ilościowa związków organicznych, Warsaw, PWN, 1979 (in Polish).

5. Czermiński J.B., Iwasiewicz A., Paszek Z., and Sikorski A., *Metody statystyczne dla chemików*, Warsaw, PWN, 1986 (in Polish).
6. Accuracy (trueness and precision) of measurement methods and results—Part 2: Basic method for the determination of repeatability and reproducibility of a standard measurement method, ISO 5725–2:1994.
7. Doerffel K., *Statystyka dla chemików analityków*, Warsaw, WNT, 1989 (in Polish).
8. Davies P.L., Statistical evaluation of interlaboratory tests, *Fresenius Z. Anal. Chem.*, 331, 513–519, 1988.
9. Linsinger T.P.J., Kandel W., Krska R., and Grasserbauer M., The influence of different evaluation techniques on the results of interlaboratory comparisons, *Accred. Qual. Assur.*, 3, 322–327, 1998.
10. Cortez L., Use of LRM in Quality Control: Interlaboratory Testing—EC Growth Projects TRAP-LRM/TRAP-NAS, 2001.
11. ISO/IEC Guide 43-1 Proficiency testing by interlaboratory comparison—Part 1: Development and operation of proficiency testing schemes.
12. Van Dyck K., Robberecht H., Van Cauwenbergh R., Deelstra H., Arnaud J., Willemyns L., Benijts F., Centeno J.A., Taylor H., Soares M.E., Bastos M.L., Ferreira M.A., D'Haese P.C., Lamberts L.V., Hoenig M., Knapp G., Lugowski S.J., Moens L., Riondato J., Van Grieken R., Claes M., Verheyen R., Clement L., and Uytterhoeven M., Spectrometric determination of silicon in food and biological samples: An interlaboratory trial, *J. Anal. At. Spectrom.*, 15, 735–743, 2000.
13. Cools N., Delanote V., Scheldeman X., Quataert P., De Vos B., and Roskams P., Quality assurance and quality control in forest soil analyses: A comparison between European soil laboratories, *Accred. Qual. Assur.*, 9, 688–694, 2004.
14. Achnazarowa S.Ł., and Kafarow W.W., *Optymalizacja eksperymentu w chemii i technologii chemicznej*, Warsaw, WNT, 1982 (in Polish).

2 Quality of Analytical Results

2.1 DEFINITIONS [1–3]

Analytical quality: Consistency of the obtained results (chemical analysis) with the accepted assumptions. The quality of information can be divided into components: quality of results, quality of the process, quality of the instruments, and quality of the work and organization.

Quality: The realization of specific requirements (which include the standards established by the quality control system in addition to accepted in-house requirements).

Quality control: A complex system of actions to obtain measurement (determination results) with the required quality level. A program of quality control includes:

- Assuring a suitable level of staff qualifications
- Assuring the proper calibration of instruments and laboratory equipment
- Good laboratory practice (GLP)
- Standard procedures

2.2 INTRODUCTION

The past decade or so was undoubtedly the period of "information hunger." Access to a variety of information sources facilitates decision making not only in politics, but also in the economy and technology (related to control over the processes of manufacturing consumer goods). A new type of market arose where information is bought and sold.

Analytical data on the researched material objects are a specific kind of information. This information is not usually obtained through an analysis of the whole object, but is based on the analyses of appropriate samples. Therefore, samples have to be collected in such a way that the most important criterion—that is, representativeness—is met.

To satisfy the growing demand for analytical data, more and more intense research is taking place with the aim of developing new methodologies and devices so that the analytical results are a source of as much information as possible—in other words, that they are characterizied by the greatest information capacity possible.

Measurement results must be reliable. That means they must accurately (both truly and precisely) reflect the real content (amount) of analytes in a sample that is representative of the material object under research. This leads to the conclusion that all developments in analytical chemistry are derived from the desire to obtain in-depth analytical data [4].

The notion of reliability is closely associated with the notion of quality. It is the quality of a result, together with its control and assurance, that determines and confirms its reliability. In analytics, the notion of quality has a specific meaning.

Results of analytical measurements are a type of product of the chemical analyst's work.

Both manufactured products and analytical results must have an appropriate quality. In addition, the quality of analytical measurements appears to have its own accumulative requirement: The quality of every product is a result of comparison of the obtained value with the reference value, expected or standard. For the obtained result to be comparable (authoritative, reliable) to the reference value, its (high) quality must be documented and maintained. The quality of results of analytical measurements must be assured in the first place to draw conclusions about the quality of the examined products.

2.3   QUALITY ASSURANCE SYSTEM

One of basic trends in the recent development of analytical chemistry is determining lower and lower concentrations of analytes in samples with a complex matrix. The need for a uniform and defined control system, of estimation and assuring the quality of analytical results, is a consequence of the following trends in analytics:

- Decrease in the concentrations of analytes
- Increase in the complexity of the matrix composition of the sample
- Introduction of new notions associated with the application of metrology principles in analytics
- Necessity of traceability documentation and estimating uncertainty as requisite parameters of an analytical result
- Globalization and the associated necessity of comparing results in different laboratories

This task poses a great challenges for analysts and draws attention to quality assurance and quality control (QA/QC) of the obtained results. The system of quality estimation usually includes the following elements:

- Tracking and estimating the precision of obtained results by periodic analysis of test samples
- Estimation of accuracy by
  - Analyses of certified reference samples
  - Comparison of obtained results with results obtained for the same sample using the reference method
  - Sample analyses after the addition of a standard
  - Comparative interlaboratory (intercomparison) exercises
- Control charts
- Suitable audit system

At present, there are three systems of quality assurance in analytical laboratories [5]:

- Good laboratory practice (GLP)
- Accreditation of a laboratory according to ISO Guide 17025 or EN 45001
- Certification according to norms ISO of series 9000

The selection of the quality system, introduced by a given laboratory, is in principle voluntary, although increasing attention is paid to the procedures of accreditation [6].

The problem relating to quality assurance and control of measurement results is primarily associated with the insufficient amount of information concerning instruments used in the process and its application. These are first of all statistical instruments based on metrology.

Quality assurance of analytical measurement results is the system comprising five interdependent elements [7]:

- Assurance of measuring traceability of the obtained results
- Estimation of uncertainty obtained in results of measurement
- Use of certified reference materials
- Participation in various interlaboratory comparisons
- Validation of the applied analytical procedures

Only when the aforementioned tools are used is it possible to provide the authoritative (reliable) results of analytical measurements.

In Figure 2.1, a schematic presentation of the elements of a quality assurance/quality control system used for obtaining reliable analytical results is shown [7].

The elements of the quality system are interdependent. To assure measuring traceability, it is indispensable to use both the certified reference materials and the analytical procedures subject to prior validation.

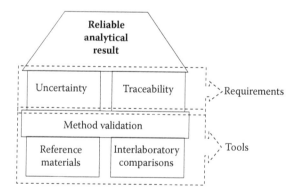

FIGURE 2.1   Position and role of the quality assurance/quality control (QA/QC) system elements for obtaining a reliable analytical result.

During the validation of an analytical procedure it is necessary to

- Use certified reference materials—determine the accuracy
- Participate in the interlaboratory comparisons—determine the traceability and reproducibility (ruggedness)
- Estimate uncertainty—which enables the control of the entire analytical procedure

Interlaboratory comparisons involve both reference materials and validated analytical procedures. On the other hand, this type of research serves to determine certified values for the manufactured reference materials.

In the production of reference materials, validated analytical procedures are applied during the determination of homogeneity and stability of materials. Reference material is also characterized by the uncertainty value.

Estimation of measurement uncertainty, as noted earlier, is indispensable in the production of reference materials.

Although the uncertainty is not one of the validation parameters, it is obvious that the determination of uncertainty increases the reliability of the obtained results. It is because during the design of the so-called "uncertainty budget" it is requisite to determine the influence of all the possible parameters of an analytical procedure on the value of the combined uncertainty. This, in turn, compels the precise and very attentive "tracking" of the entire analytical procedure, those enabling the control of the procedure.

Interrelations among the particular QA/QC system components are presented in Figure 2.2.

Each element of the quality control system concerning the results of analytical measurements must be applied by any laboratory that wishes to obtain reliable results. Each of these elements, in order to be applicable, must be defined in a way that is intelligible for the user. Its must also be clearly and intelligibly presented, along with the determination and the control of the elements of the quality control. This can be achieved by

- Defining the basic notions of the quality system
- Determining the simple and intelligible procedures used when using individual elements of the quality system
- Providing clear and transparent dependencies (in which elements of metrology and mathematical statistics are used) enabling the "numerical" or "parametric" determination of each characteristic, and the determination of quality of the control system elements
- Helping users to derive inferences on the quality system, based on determined values for each of its elements

Every analyst should be aware that the basic and requisite parameters characterizing an analytical result are traceability and uncertainty. These two parameters are the basic requirements for a reliable measurement result. A schematic representation of this concept is shown in Figure 2.3.

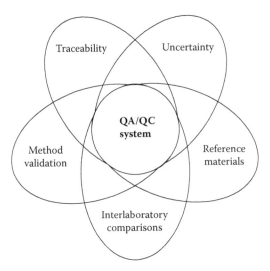

FIGURE 2.2 Components of the QA/QC system of an analytical process, showing inter-relationships among components.

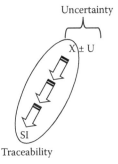

FIGURE 2.3 Necessary parameters for a reliable analytical result.

The necessity of presenting the result together with these two basic parameters must be remembered by every "producer" of an analytical result.

A requisite condition of assuring the appropriate quality of analytical results is verification of the reliability of the used gauges and checking of the range of applica-tion and calibration of the analytical procedures. Accordingly, analytical procedures usually involve two operations associated with calibration:

- Periodic reliability test of indications of the instruments used by means of standard mixtures; a special case of such mixtures is "zero" mixtures used for
  - Testing the zero position on the measuring scale of the instrument
  - Diluteness of standard mixtures, containing strictly defined concentra-tions of analytes
- Testing the reliability of the whole plot of the analytical conduct

Realization of this operation can be achieved in two ways:

- By addition of a standard to the analyzed sample
- As a result of applying reference material samples

Chemical analysis of any material can be described as a chain of decisions, actions, and procedures [8]. As in the case of any chain, also in a chemical analysis the power of the entire chain depends on the power of its weakest link. In general, the weakest links in the analytical process are not the elements acknowledged as components of chemical analysis (e.g., chromatographic extraction of mixtures or spectrometric detection), but rather the stages that take place outside the analytical laboratory, such as

- Selection of materials to be sampled
- Preparation of the sampling strategy
- Selection and use of techniques and devices necessary in sampling, and also their transport, maintenance, and storage

If a given analytical laboratory is not responsible for the sampling stage, the quality management system does not take into account these weak steps of the analytical process. Moreover, if stages of sample preparation (extraction, purifying extracts) have not been carried out properly, then even the most modern analytical instruments and complex computer techniques cannot improve the situation. Such analytical results have no value and instead of being a source of information, can cause serious misinformation. Hence, QA/QC of the analytical results should involve all stages of the analytical process. This process must be an integral, where the applicability test for the analytical method (validation) is only one stage, albeit an important one.

2.4 CONCLUSION

For a laboratory to be able to deliver reliable and repeatable results, it is necessary to perform systematic calibration of analytical instruments and subject all analytical procedures to validation. This notion means the determination of the methodology characteristics, covering the previous notion of "method applicability range" (selectivity, accuracy, precision, repeatability, limit of detection, range, linearity, etc.). For the purpose of quality control in laboratory work, reference material samples are subject to the same processing and determinations as real samples. Comparison of the obtained result, with the real analyte concentration in the reference material sample, may give conclusions concerning the reliability of analytical works conducted in a given laboratory [9].

REFERENCES

1. Konieczka P., and Namieśnik J. (eds.), *Kontrola i zapewnienie jakości wyników pomiarów analitycznych*, WNT, Warsaw, 2007 (in Polish).

2. Hulanicki A., Współczesna chemia analityczna. Wybrane zagadnienia, PWN, Warsaw, 2001 (in Polish).
3. Paneva V.I., and Ponomareva O.B., Quality assurance in analytical measurements, *Accred. Qual. Assur.*, 4, 177–184, 1999.
4. Richter W., How to achieve international comparability for chemical measurements, *Accred. Qual. Assur.*, 5, 418–422, 2000.
5. Hembeck H.-W., GLP and other quality assurance systems—A comparison, *Accred. Qual. Assur.*, 7, 266–268, 2002.
6. Dobecki M. (ed.), Zapewnienie jakości analiz chemicznych, Instytut Medycyny Pracy im. Prof. J. Nofera, Łódź, 2004 (in Polish).
7. Konieczka P., The role of and place of method validation in the quality assurance and quality control (QA/QC) system, *Crit. Rev. Anal. Chem.*, 37, 173–190, 2007.
8. Valcárcel M., and Rios A., Analytical chemistry and quality, *Trends Anal. Chem.*, 13, 17–23, 1994.
9. Mermet J.M., Otto M., and Valcárcel M. (eds.), *Analytical Chemistry: A Modern Approach to Analytical Science*, Weinheim: Wiley-VCH, 2004.

3 Internal Quality Control

3.1 DEFINITION

Internal quality control (IQC): The set of procedures undertaken by laboratory staff for the continuous monitoring of operation and the results of measurements in order to decide whether results are reliable enough to be released.

3.2 INTRODUCTION

Internal quality control (IQC) is extremely important to ensure that the data released from the lab are fit for purpose. Quality control (QC) methods are enabled to monitor the quality of the data produced by the laboratory on a run-by-run basis [1–3].

The laboratory should run the control samples together with the routine samples. The control values are plotted in a control chart. In this way it is possible to demonstrate that the measurement procedure executes within the given limits. If the obtained value is outside the control limits, no analytical results are reported and corrective actions must be taken to identify the error sources, and to remove such errors.

If the laboratory is accredited, the standard ISO/IEC 17025 requires that the laboratory assesses the needs of the user, before any analysis. Each laboratory should define its quality requirements [4].

IQC takes place within the analytical series or runs. The main purpose of IQC is to answer the question: Does my method consistently fit my purpose?

Once a laboratory has implemented a method in its routine work, is performing adequate QC, has taken any appropriate corrective and/or preventive actions, and its staff has acquired sufficient expertise, it may consider including this method in its scope of accreditation [5–8].

3.3 QUALITY CONTROL IN THE LABORATORY

Quality in the laboratory can be controlled on the two levels: internally and externally. Both are characterized schematically in Figure 3.1.

The main objectives of QC in the laboratory [9] are to

- Help lab staff establish, manage, and monitor a testing process to assure the analytical quality of test results
- Determine problem and solve them
- Develop uniform standards of laboratory
- Increase lab staff and client confidence
- Create good databases for decisions makers
- Reduce cost

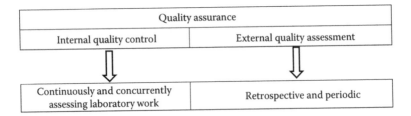

FIGURE 3.1   Internal QC and external quality assessment as a part of QA/QC system.

On the other hand, the main goals of QC are to

- Detect significant errors rapidly
- Report good results in a timely manner
- Be cost-effective and simple to use
- Identify sources of errors when they occur

There are lots of analytical factors that can influence quality. The majority of them include [10] the following:

- Proficiency of the personnel: Education, training, competence, commitment, adequate number, supervision, motivation.
- Reagent stability, integrity, and efficiency: Stable, efficient, desired quality, continuously available, checked (e.g., purity).
- Equipment reliability: Meet technical needs; compatible; user- and maintenance-friendly; cost-effective; validated (known value of metrological parameters); adequate space; storage and segregation for incompatible activities; controlled and monitored environmental conditions; suitable location; "fit for purpose"—validation, calibration—documented program; maintenance; records.
- Selectivity and sensitivity of selected procedures—validated, standard operating procedure, including every step of the analytical procedure.
- Use of appropriate controls.
- Use of appropriate recording and documentation including all written policies, plans, procedures, instructions, records, quality control procedures, and recorded test results involved in providing a service or the manufacture of a product.

The IQC must be planned to avoid unnecessary activities on the one hand and not taking into account all the parameters on the other. The planning of QC has to include [9]

- Checking the appropriate concentrations and types of control samples according to the scope of the laboratory's method.
- Definition of the purpose of each control: Whole method, part of method (e.g., control of calibration drift).

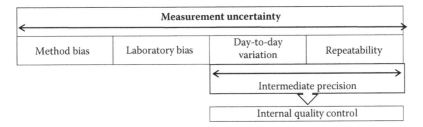

FIGURE 3.2 The place of IQC in uncertainty budget preparation.

- Taking into account control at the beginning and end of each series.
- Intermediate checking of the stability of the measurement process and stability of samples.
- Selection of what goes into the database for the generation of updates on general quality of analyses (when plotting results).

Data obtained during IQC can be used (and should be used) as an element of uncertainty budget for the procedure. It is mainly used as an uncertainty arising from precision, as is presented schematically in Figure 3.2.

During IQC several tools can be used:

- Standards
- Standard solutions
- Blank samples
- Appropriate certified reference material (CRM)
- Fortified samples
- Statistical treatment—mainly control charts

3.4 CONTROL CHARTS

3.4.1 Shewhart Charts

Control charts are used to test the stability of research results conducted in a given laboratory. In practice, the most frequently used charts are Shewhart charts. This method of monitoring and regulating processes is a graphic procedure minimizing the number of necessary numerical operations and allowing systematic monitoring of the course of the process being subjected to control. It enables fast and simple detection of abnormalities in the configuration of the marked points, and thus fast correction and confirmation of the reliability of the research [11–14].

The main role is played by an appropriate control chart, usually a graph with control limits depicted. On such a graph, the values of a certain statistics measurement are registered. The measurement is obtained from a series of measurement results obtained at approximately regular intervals, expressed either in time (e.g., every hour) or quantity (e.g., every batch).

There are two types of variability in the charts:

- Variability due to random change
- Real variability of the parameter in the process

3.4.2 SHEWHART CHART PREPARATION

Preparation of a chart depicting mean and standard deviation $(x_m - SD)$ will be described as an example; charts are prepared separately for each procedure [12–16].

The course of action in preparing the control chart is as follows:

- Conduct 10–20 measurements for a standard sample.
- Calculate the mean x_m and the standard deviation SD; both values should be determined for the unbiased series, that is, after the initial rejection of outliers.
- Test the hypothesis about a statistically insignificant difference between the obtained mean and the expected value using Student's t test (Section 1.8.9).
- If the hypothesis is not rejected, start the preparation of the first chart:
  - Mark the consecutive numbers of result determinations on the x-axis of the graph, and the values of the observed characteristics (the mean) on the y-axis.
  - Mark a central line (CL) on the graph corresponding to the reference values of the presented characteristic, and two statistically determined control limits, one line on either side of the central line; the upper and lower control limits (UAL and LAL, respectively), or in other words the upper and lower warning limits. Both the upper and lower limits on the chart are found within $\pm 3 \times SD$ from the central line, where SD is the standard deviation of the investigated characteristics. Limits of $\pm 3 \times SD$ (so-called action limits) show that approximately 99.7 percent of the values fall in the area bounded by the control lines, provided that the process is statistically ordered. The possibility of transgressing the control limits as a result of random incident is insignificantly small; hence, when a point appears outside the control limits $\pm 3 \times SD$, it is recommended that action be taken on the chart. Limits of $\pm 2 \times SD$ are also marked; however, the occurrence of any value from a sample falling outside these limits is simply warning about a possible transgression of the control limits; therefore, the limits of $\pm 2 \times SD$ are called warning limits (UWL and LWL).
  - Mark the obtained measurement results for 20 consecutive samples, but the results obtained for control samples should be marked parallel to the received results for the investigated samples:

- If a determination result is located within the warning limits, it is considered satisfactory.
- The occurrence of results between the warning limits and action limits is also acceptable; however, not more often than two results per twenty determinations.
- If a result for a test sample is found outside the action limits, or seven consecutive results create a trend (decreasing or increasing), calibration should be carried out again.
- There exist three other signs indicating the occurrence of a problem in the analyzed arrangement, namely:
  - Three consecutive measurement points occurring outside the warning limits, but within the action limits.
  - Two subsequent measurement points being outside warning limits, but in the interval determined by the action limits, on the same side of the mean value.
  - Ten consecutive measurement points being found on the same side of the mean value.

3.4.3 SHEWHART CHART ANALYSIS

For each new chart, it is necessary to compare the mean obtained for test samples with the expected value. When the difference between these values is statistically significant, the results from this series (chart) should be rejected. Otherwise, one should compare the standard deviations obtained for the investigated chart and those obtained for a previous chart using Snedecor's F test (Section 1.8.5), albeit the comparison should always involve two last charts. If the standard deviations do not differ in a statistically significant manner, the standard deviation is calculated for the next chart as the square root of arithmetic mean of the variance (V) values for the compared charts.

Compare the mean values obtained for the investigated chart and those obtained for a previous chart using Student's t test (Section 1.8.9). If the means do not differ in a statistically significant manner, a new mean is calculated for the next chart as the arithmetic mean for the compared chart, and a new chart is prepared for the newly calculated values of x_m and SD.

If the standard deviations differ in a statistically significant manner, a new chart should be prepared for the values of the penultimate chart.

If the mean values differ only in a statistically significant manner, a new chart should be prepared for parameters identical to those in the compared chart.

If the process is statistically regulated, then a control chart is the method used for continuously testing the statistical null hypothesis and for testing whether the process is not changing and remains statistically regulated. If a value marked on the chart falls outside any of the control limits or the series of values reflects unusual configurations, the process is not statistically regulated. In this situation, one should detect the cause, and the process may then be halted or corrected. Once the cause has been located and eliminated, the process may be resumed and continued.

Example 3.1

Problem: Draw a Shewhart chart for the 20 given measurement results obtained for the test samples. Mark the central line, and the warning and action lines.
Data: Series results:

	Data
1	4.21
2	4.23
3	4.30
4	4.32
5	4.11
6	4.04
7	4.27
8	4.20
9	4.07
10	4.32
11	4.12
12	4.22
13	4.23
14	4.36
15	4.10
16	4.04
17	4.14
18	4.17
19	4.34
20	4.22

Solution:

mean	4.20
SD	0.100

Before calculating limit values, it is necessary to check if there are any outliers in the series. Equation 1.20 (Section 1.8.1) has been used, as the number of results is >10.

For $\alpha = 0.05$, $k_a = 1.65$.

The calculated interval is equal to 4.036–4.365.

All results are within the interval so there are no outliers in the series.

Calculated limits values:

UAL	4.50
UWL	4.40
LWL	4.00
LAL	3.90

Graph:

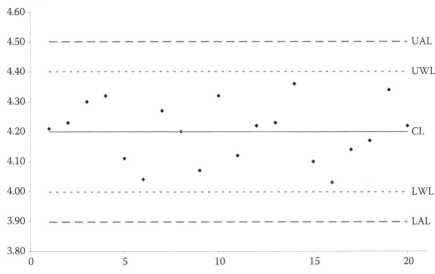

Excel file: exampl_IQC01.xls

Example 3.2

Problem: Mark the following data from the previous example on the chart.
Data: Series results:

	Data	
1	4.44	!
2	4.35	ok
3	4.12	ok
4	4.32	ok
5	4.18	ok
6	4.08	ok
7	4.34	ok
8	4.41	ok
9	4.23	ok
10	4.01	ok
11	4.11	ok
12	4.33	ok
13	4.20	ok
14	4.15	ok
15	4.17	ok
16	4.32	ok
17	4.00	ok
18	4.12	ok
19	4.11	ok
20	4.11	ok

Solution:

$mean_{\_series2}$	4.21
$SD_{\_series2}$	0.130

Before calculating limit values it is necessary to check if there are any outliers in the series. Equation 1.20 (Section 1.8.1) has been used, as the number of results is >10.

For $\alpha = 0.05$ $k_a = 1.65$.

The calculated interval is equal to 3.990–4.420.

The first result in the series is an outlier, so it has to be removed from the series, and new values of mean and SD have to be calculated.

$mean_2$	4.19
SD_2	0.121

Graph:

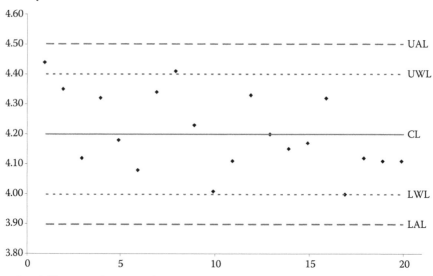

Excel file: exampl_IQC01.xls

Example 3.3

Problem: Draw a new chart based on the data from the previous example.

Solution: Value 1 has been removed from the set of data. The remaining values were used to calculate the mean and the standard deviation.

The variances were compared using Snedecor's F test, and then (with variances not differing in a statistically significant way) the means was compared using Student's t test.

	Series 1	Series 2
No results—n	20	19
standard deviation—SD	0.101	0.121
mean	4.20	4.19
F		1.472
$F_{crit(0.05,\ 18,\ 19)}$		2.182
$F < F_{crit}$		
t		0.222
$t_{crit(0.05,\ 37)}$		2.026
$t < t_{crit}$		

F-Test Two Sample Variances

	Variable 1	Variable 2
Mean	4.19	4.20
Variance	0.0146	0.00993
Observations	19	20
df	18	19
F	1.472	
$P(F \leq f)$ one-tail	0.205	
F critical one-tail	2.182	

t-Test: Two Sample Assuming Equal Variances

	Variable 1	Variable 2
Mean	4.20	4.19
Variance	0.00993	0.0146
Observations	20	19
Pooled Variance	0.0122	
Hypothesized Mean Difference	0	
df	37	
t Stat	0.222	
$P(T \leq t)$ two-tail	0.825	
t critical two-tail	2.026	

For the new chart, the values have been calculated as the means of the two previous charts.

mean	4.20
SD	0.111
UAL	4.53
UWL	4.42
LWL	3.97
LAL	3.86

Graph:

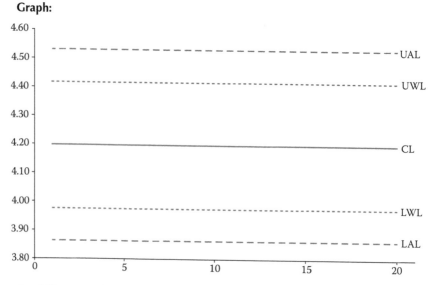

Excel file: exampl_IQC01.xls

There are ten out-of-control situations (not possible from the probability point of view). The first four are called the Western Electric Rules [12]. The ten are as follows:

1. One or more points plot outside the control limits (3-sigma limits).
2. Two out of the three consecutive points outside the 2-sigma warning limits but still inside the control limits.
3. Four of five consecutive points beyond the 1-sigma limit.
4. A run of eight consecutive points on one side of the center.
5. Six points in a row steadily increasing or decreasing.
6. Fifteen points in a row in zone 1-sigma limits (both above and below the central line).
7. Fourteen points in a row alternating up and down.
8. Eight points in a row in both sides of the central line with none in zone 1-sigma limits.
9. An unusual or non-random pattern in the data.
10. One or more points near a warning or control limit.

For each new chart it is necessary to compare the mean obtained for test samples with the expected value. When the difference between these values is statistically significant, the results from this series (chart) should be rejected. Otherwise, one should compare the standard deviations obtained for the investigated chart and those obtained for a previous chart using Snedecor's F test, albeit the comparison should always involve the two last charts, $(n - 1)$ and (n), where

$(n - 1)$: Parameters for the charts
(n): Parameters calculated from the set of data

If the standard deviations do not differ in a statistically significant manner, the standard deviation is calculated for the next chart as

$$SD_{n+1} = \sqrt{\frac{SD_{n-1}^2 + SD_n^2}{2}} \qquad (3.1)$$

The mean values are compared using Student's t test.

- If the means do not differ in a statistically significant manner, a new mean is calculated for the next chart as the arithmetic mean for the compared chart, and a new chart is prepared for the newly calculated values of x_m and SD.
  SD and x_m are calculated based on charts $(n - 1)$ and (n).
- If the standard deviations differ in a statistically significant manner, a new chart should be prepared for the values of the preultimate chart.
  SD and x_m are calculated based on chart $(n - 1)$.
- If the mean values differed only in a statistically significant manner, a new chart should be prepared for parameters identical to those in the compared chart.
  SD and x_m are calculated based on chart (n).

3.4.4 TYPES OF CONTROL CHARTS

Depending on control sample used, parameter, what is to be controlled, and type of measurement, there are different types of control charts that can be used [12,17–19]:

- X-chart: An original Shewhart chart with single values used mainly for precision checks. It can be used for trueness control, though synthetic samples with known content or RM/CRM samples may be analyzed. It can be used for calibration checking (slope, intercept stability), too.
- Blank value chart: It is a special form of the X-chart, which can be used for the control contamination of reagents, state (stability, selectivity) of the analytical system and contamination sources; the conclusions are made based on direct measurements of signals, not calculated values.

- *Recovery chart*: Applied for controlling an influence of the sample matrix for recovery, it is calculated as

$$\%R = \left(\frac{x_{spiked} - x_{unspiked}}{\Delta x_{expected}} \right)[\%] \qquad (3.2)$$

with a target value around 100 percent.
- *Range chart (R-chart)*: The calculated parameter is an absolute difference between the highest and lowest value of multiple analyses, it can be applied for different analyte content—then the relative value can be used. This control chart has the only upper limits.

Example 3.4

Problem: For given series of data calculate R-chart parameters. Make calculations for range and relative range as well.
Data: Series results:

No.	Date	Result 1	Result 2
1	17.12.09	760	751
2	19.02.10	596	604
3	30.03.10	703	693
4	18.08.10	4706	4718
5	30.09.11	36	36.8
6	20.01.12	37.7	37.1
7	27.01.12	4205	4192
8	10.02.12	924	930
9	15.02.12	7826	7859
10	24.02.12	478	490
11	27.02.12	836	820
12	16.03.12	32	31.5
13	30.03.12	793	803
14	27.04.12	687	675
15	12.06.12	6717	6693
16	13.06.12	32.7	33.4
17	14.06.12	17.5	17.9
18	20.07.12	45	46.1
19	17.08.12	28.5	28.3
20	22.08.12	6887	6850

Solution:
Range calculated for all results as |*result 1 – result 2*|.

No	R	Conclusion
1	9.0	+
2	8.0	+
3	10.0	+
4	12.0	+
5	0.8	+
6	0.6	+
7	13.0	+
8	6.0	+
9	33.0	+
10	12.0	+
11	16.0	+
12	0.5	+
13	10.0	+
14	12.0	+
15	24.0	+
16	0.7	+
17	0.4	+
18	1.1	+
19	0.2	+
20	37.0	-

Calculated limits values:

D4	3.267
UCL	33.7
CL	10.3

where CL = average value of range

$UCL = D4 \times CL$

Conclusion: Based on the limits values calculated for range, the results in row 20 are out of the UCL.

Calculation of relative range as range/average:

No	R_{rel}	Conclusion
1	1.2%	+
2	1.3%	+
3	1.4%	+
4	0.3%	+
5	2.2%	+
6	1.6%	+
7	0.3%	+

8	0.6%	+
9	0.4%	+
10	2.5%	+
11	1.9%	+
12	1.6%	+
13	1.3%	+
14	1.8%	+
15	0.4%	+
16	2.1%	+
17	2.3%	+
18	2.4%	+
19	0.7%	+
20	0.5%	+

D4	3.267
UCL	4.4%
CL	1.3%

Conclusion: Based on limits values calculated for relative range no value out of the UCL.

Due to different analyte content the correct way of calculations is that one that used relative range.

Excel file: exampl_IQC02.xls

- *CUSUM chart (CUMulative SUM)*: It is a highly sophisticated control chart, the CUSUM is a sum of all differences from one target value, which value is subtracted from every control analyses and difference added to the sum of all previous differences, the recognition of a systematic change in the mean value is very simple, there is possible to determine of the order of magnitude by which the mean value has changed and the point in time at which the change occurred; the main advantages of that chart are [11,14]
  - An indication at what point the process went out of control
  - The average run length is shorter
  - The number of points that have to be plotted before a change in the process mean is detected
  - The size of a change in the process mean estimated from the average slope

  Reference value (*k*) could be either an assigned value (CRM, RM, spiked sample) or the value determined in the preliminary period; the standard deviation is determined in the training period, the V-mask is the base of a two-sided statistical test and is defined by the following parameters:
  - *d*, expressed in abscissa units, is the distance from the vertex of the mask to the more recent entry on the chart
  - *θ* is the angle between the arms of the mask and the horizontal drawn through the mask vertex

The V-mask is usually drawn on a transparent film. It is positioned on the CUSUM chart with the vertex at a distance d from the latest entry; thus, for each new entry, the mask is shifted one abscissa unit parallel to the time axis; an out-of-control situation is indicated if the CUSUM line crosses one of the arms of the V-mask. If the CUSUM line cuts the upper arm, then the mean value has decreased and vice versa; the first CUSUM value that lies outside of the mask indicates the point in time at which the out-of-control situation appeared; this information can be helpful when searching for the cause of error; the larger the values of d and θ, the more infrequently an out-of-control situation arises.

Example 3.5

Problem: Draw a Shewhart chart for the 20 given measurement results obtained for the test samples. Mark the central line, and the warning and action lines.
    Calculate also data for CUSUM chart and make an appropriate graph.
Data: Series results:

	Data
1	42
2	44
3	43
4	42
5	44
6	41
7	44
8	42
9	40
10	41
11	38
12	39
13	40
14	42
15	41
16	40
17	38
18	38
19	39
20	41

Target	42
SD	3

Solution:
Before calculating the limit values it is necessary to check if there are any outliers in the series. Equation 1.20 (Section 1.8.1) has been used, as the number of results is >10.
For $\alpha = 0.05$ $k_\alpha = 1.65$.
Calculated interval is equal 37.7–44.2.
All results are within the interval so there are no outliers in the series.
Calculated limits values:

Mean	41.0
SD	2.0
UAL	46.8
UWL	44.9
LWL	37.0
LAL	35.1

Graph:

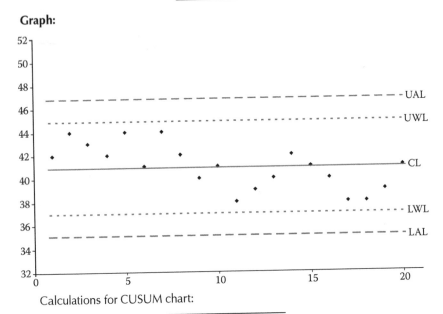

Calculations for CUSUM chart:

	Data	CUSUM
1	42	0
2	44	2
3	43	3
4	42	3
5	44	5
6	41	4
7	44	6
8	42	6
9	40	4
10	41	3
11	38	−1
12	39	−4

13	40	−6
14	42	−6
15	41	−7
16	40	−9
17	38	−13
18	38	−17
19	39	−20
20	41	−21

Graph:

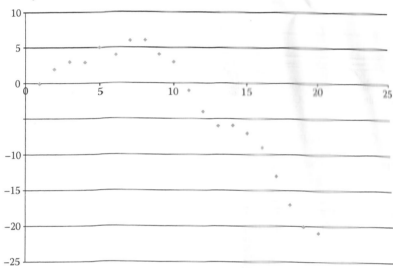

After putting on V-mask:

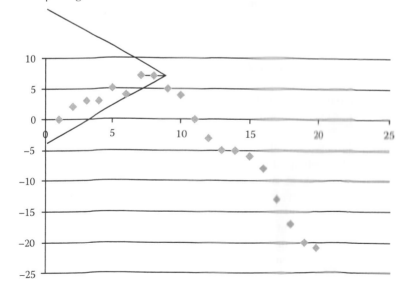

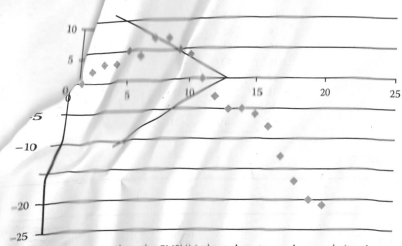

This is a moment when the CUSUM chart detects an abnormal situation.
Excel file: exampl_IQC03.xls

3.4.5 CONTROL SAMPLES

Appropriate control samples, used for control charts, have to fulfill the following requirements [11,14]:

- Be representative for matrix and analyte concentration and concentration in the region of analytically important values (limits!)
- Be homogeneous
- Be stable for at least several months under defined storage conditions
  - Regular removal of sample aliquots for the control analyses must not lead to changes in the control sample
  - Should have enough available

The use of control samples must be decided, taking into account a compromise/balance situation between the cost and time, and the risk to undetected analytical errors.

In order to avoid the effects of an unknown cyclicity or to detect them by applying different types of control samples, it is possible to control different parameters:

- Control Samples—standards: It can be used to verify the calibration, but the control sample must be completely independent from calibration solutions, the influence of sample matrix cannot be detected; the control precision and trueness (no matrix effect) is limited.
- Control Samples—blank: It can detect errors due to changes in reagents, in new batches of reagent carryover errors, and in drift of apparatus parameters; blank samples analyzed at the start and at the end allow identification of some systematic trends.

TABLE 3.1

Suitability of Different Types of Control Samples

Sample Type	Trueness	Precision
CRM	Yes	Yes
QCM (RM)	Yes (CRM)	Yes
PT sample	Yes	Yes
Real sample	No	Yes
Spiked real sample	Yes (% Recovery)	Yes
Blank sample	Yes (blank value)	Yes (blank value)
Synthetic sample	Yes (if representative)	Yes (if representative)
Standard solution	Yes (calibration)	Yes (calibration)

- Control Samples—real samples: They can be used for multiple analyses of range and differences charts if it is necessary to separate charts for different matrices; they can be used for rapid precision control, but not for trueness checks.
- Control Samples—RM, CRM: These are ideal control samples, but they are too expensive or unavailable for all types of analysis; in-house reference materials are a good alternative. One can be checked regularly against a CRM; the retained sample material from interlaboratory tests can be used.

General information about suitability of different control samples is presented in Table 3.1.

It is necessary to point out that the more frequently a specific analysis is done, the more sense a control chart makes. If the analyses are always done with the same sample matrix, the sample preparation should be included. If the sample matrix varies, the control chart can be limited to the measurement only.

3.5 CONCLUSION

IQC in a chemical analytical laboratory is the continuous, critical assessment of a laboratory's own analytical methods and procedures. This control includes the analytical process, starting with the sample entering the laboratory and finishing with an analytical report. The most important tool in QC is the use of control charts. The basis for this is the laboratory work control samples together with the routine samples. The results of control can be used in several ways—the analyst will have a very important tool in his or her daily work; the client may get the impression that laboratory quality and results can be used in the estimation of the uncertainty of measurement. IQC is to be a part of the quality system and should be formally reviewed at regular intervals. That control could be treated as a continuous process during the operational lifetime of an analytical method, while validation is a periodic one. Schematically it is presented in Figure 3.3.

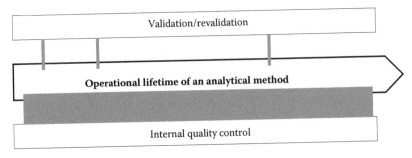

FIGURE 3.3 Frequency of IQC and method validation during the operational lifetime of an analytical method.

REFERENCES

1. Massart D.L., Vandeginste B.G.M., Buydens L.M.C., De Jong S., Lewi P.J., and Smeyers-Verbeque J., *Handbook of Chemometrics and Qualimetrics*, Elsevier, 1997.
2. Mullins E., *Statistics for the Quality Control Chemistry Laboratory*, The Royal Society of Chemistry, 2003.
3. Wenclawiak B.W., Koch M., and Hadjicostas E. (Eds.), *Quality Assurance in Analytical Chemistry*, Springer, 2003.
4. Funk W., Dammann V., and Donnevert G., *Quality Assurance in Analytical Chemistry*, Second Edition, VCH, 2007.
5. Thompson M., and Lowthian P.J., Effectiveness of analytical quality control is related to the subsequent performance of laboratories in proficiency tests, *Analyst*, 118, 1495–1500, 1993.
6. Royal Society of Chemistry, Analytical Methods Committee, Internal quality control of analytical data, *Analyst*, 120, 29–34, 1995.
7. Thompson M., and Wood R., Harmonized guidelines for internal quality control in analytical chemistry laboratories, *Pure Appl. Chem.* 67, 649–666, 1995.
8. Gardner M.J., Quality control techniques for chemical analysis: Some current shortcomings and possible future developments, *Accred. Qual. Assur.* 12, 653–657, 2007.
9. Nordtest Technical Report. *Internal Quality Control. Handbook for Chemical laboratories.* Available at http://www.nordtest.info/index.php/technical-reports/item /internal-quality-control-handbook-for-chemical-laboratories-trollboken-troll-book-nt -tr-569-english-edition-4.html (accessed August 9, 2017).
10. Thomson M. (ed.), Internal quality control in routine analysis. AMC technical briefs No 46. RSC 2010. Available at http://www.rsc.org/images/internal-quality-control-routine -analysis-technical-brief-46_tcm18-214818.pdf (accessed August 9, 2017).
11. Mullins E., Introduction to control charts in the analytical laboratory. Tutorial review, *Analyst*, 119, 369–375, 1994.
12. Howarth R.J., Quality control charting for the analytical laboratory: Part 1. Univariate methods: A review, *Analyst*, 120, 1851–1873, 1995.
13. Mestek O., Pavlík J., and Suchánek M., Robustness testing of control charts employed by analytical laboratories, *Accred. Qual. Assur.* 2, 238–242, 1997.
14. Mullins E., Getting more from your laboratory control charts. Tutorial review, *Analyst*, 124, 433–442, 1999.
15. ISO 7870-1:2014. Control charts. Part 1: General guidelines.
16. ISO 7870-2:2013. Control charts. Part 2: Shewhart control charts.
17. ISO 7870-3:2012. Control charts. Part 3: Acceptance control charts.
18. ISO 7870-4:2011. Control charts. Part 4: Cumulative sum charts.
19. ISO 7873:1993. Control charts for arithmetic average with warning limits.

4 Traceability

4.1 DEFINITIONS [1]

International (measurement) standard: Measurement standard recognized by signatories to an international agreement and intended to serve worldwide.

Measurand: Quantity intended to be measured.

Measurement standard (etalon): Realization of the definition of a given quantity, with stated quantity value and associated measurement uncertainty, used as a reference.

National (measurement) standard: Measurement standard recognized by national authority to serve in a state or economy as the basis for assigning quantity values to other measurement standards for the kind of quantity concerned.

Primary standard: Measurement standard established using a primary reference measurement procedure, or created as an artifact, chosen by convention.

Reference standard: Measurement standard designated for the calibration of other measurement standards for quantities of a given type in a given organization or at a given location.

Secondary standard: Measurement standard established through calibration with respect to a primary measurement standard for a quantity of the same type.

Traceability: Property of a measurement result whereby the result can be related to a reference through a documented unbroken chain of calibrations, each contributing to the measurement uncertainty.

Traveling standard: Measurement standard, sometimes of special construction, intended for transport between different locations.

Working standard: Measurement standard that is used routinely to calibrate or verify measuring instruments or measuring systems.

4.2 INTRODUCTION

Comparison of measurement results is sensible only when they are expressed in the same units or on the same scale. The problem of traceability appeared with the first measurements carried out by man. However, the notion of traceability itself was formulated much later, in association with the development of metrological infrastructure, initially in reference to measurements of physical properties, but later with relation to chemical measurements [2].

In the handbook *Quality Management and Quality Assurance Vocabulary*, ISO 8402 [3], traceability is defined as "the ability to verify the history, location, or application of an item by means of recorded identification."

According to the International Vocabulary of Basic General Terms in Metrology (VIM) [1], traceability is defined not only as a property of a measurement result, but also as a property of a reference standard. In general, it can be described as a continuous and logical process which discourages weak or missing activity at any step of an analytic process, which could burden or lower the effectiveness of the entire process.

Every day throughout the world, millions of chemical analyses are carried out, and each has its own requirements concerning the quality of an obtained result [4]. The obtained measurement results should be traceable to respective international standards [5,6]. For example, for mass determination, a balance should be used which is calibrated regularly via weights with a calibration certificate that describes a reference to higher-order standard weights. These, in turn, should be calibrated against the national standards related to the international prototype kilogram. Such a series of comparisons is an uninterrupted chain illustrating the very property of traceability. Knowing uncertainty values at each step of this chain of comparisons, one can qualify the uncertainty of the value measured.

The schematic presentation of traceability meaning is shown in Figure 4.1 [7], while the rationale and meaning behind it is presented in Figure 4.2.

The traceability could be achieved by comparison of result value with [2]:

- SI unit
- Value represented by well-stated standard
- Value obtained by primary (absolute) method
- Value obtained by reference (excellence) laboratory
- Value obtained by group of laboratories in systematic PT scheme

It should be noted that the value of the result is traceable not to the reference material (RM) or primary method, but to the value (property) represented by RM or produced using primary method [8].

| 1 kg mass standard (Sevres, France) | Official copy of mass standard | National mass standard (1 kg) | Mass measurement |

FIGURE 4.1 The idea of traceability: An example of mass determination. (From De Bièvre, P., and Taylor, P.D.P., *Metrologia*, 34, 67–75, 1997.)

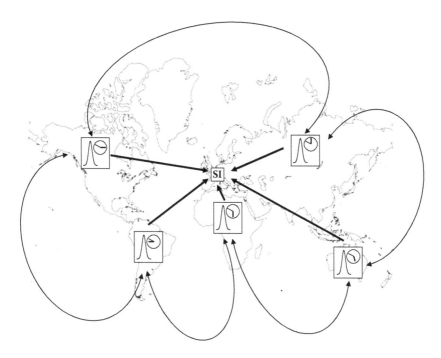

FIGURE 4.2   Rationale and meaning of traceability.

4.3   THE ROLE OF TRACEABILITY IN QUALITY ASSURANCE/QUALITY CONTROL SYSTEMS

The accuracy of an analytical result depends directly on the material used for calibration. For chemical measurements, if the determined substance is available as a certified reference material, it can be treated as the last cell of an uninterrupted chain of comparisons, that is, traceability. Thus, the most important feature of reliable measurement result is its traceability in relation to the recognized standard with well-known metrological characteristics. Assurance of traceability is realized by comparing given properties to a higher-order standard. In compliance with the content of VIM [1]: "Traceability is a property of the result of a measurement or the value of a standard whereby it can be related to stated references, usually national or international standards, through an unbroken chain of comparisons all having stated uncertainties." The quoted definition of traceability underlines the elements which are especially significant in chemical measurements. Traceability is primarily a feature of a result of measurement obtained with the use of a given measurement procedure. This result must be related to a reference standard, so that it may be expressed in suitable units. Moreover, the connection should be realized by means of an uninterrupted chain of comparisons, and at every step uncertainty should be defined. A critical step of assuring traceability in chemical measurements is the applied analytical procedure, because every physical or chemical operation can fracture this chain.

In compliance with the requirements of metrology, that is, the science of measurement, traceability is one of the most important elements of the quality of a result. Because results of measurements of physical and chemical properties are the basis of many decisions, respective scientific and legal metrology centers have emerged on an international scale.

In measuring physical properties, the result of a measurement depends substantially on the quality of the measuring instruments (rule, thermometer, scale) used, and in principle does not depend on the type of examined object. In measuring chemical values, apart from the scale calibration of a gauge, the result of measurement depends to a significant degree on the type of the sample and how the analytical procedure is conducted. Chemical measurements usually require sample preparation steps, which means the necessity of obtaining a representative dose of the examined material, and, for example, dissolution or mineralization of a sample, enrichment, and extraction (just to mention the most important physicochemical processes).

Therefore, in chemical measurements, the notion of accuracy is difficult to define, and proving traceability is considerably more difficult. In the case of chemical measurements, there is no organized metrological system similar to physical measurements realized by a system of standardizing laboratories. In chemical measurements, the calibration of instruments is not a significant source of problems. The greatest problem is assuring the traceability of the entire analytical process. As it has been noted earlier, the chain of connections with standards is always broken when a sample is physically or chemically modified in the analytical process. For this reason, an extremely important element that assures the quality of chemical measurement results is the validation of the entire measurement procedure and the estimated influence of sample components on the ultimate result of the measurement.

Traceability determination in chemical measurements is associated with many difficulties resulting from the need for sample preparation before the measurement process itself. The most important difficulties are

- Identifying the object of measurement (object of determination)
- Interferences
- Homogeneity of a sample (heterogeneity of composition)
- Persistence of the sample
- Sample preparation
- Correctness of measurement realization
- Determination of uncertainty

Determination of the measurand is a crucial element in the selection of an analytical procedure. In most cases, the value of the measurand depends on the applied methodology or on the measurement conditions. Thus the results can only be compared in the same measurement conditions.

Interferences, that is, the influence of sample components on the analytical signal, depend on the type of determined substance (analyte) and the type of matrix of the sample. As noted earlier, when measuring physical properties, the sample type does not have a significant influence on the measurement result. In chemical measurements, a result of determination depends on the components accompanying the analyte. For example, it is well known that a type of acid mixture used for mineralization influences the atomic absorption signal. The type of acid affects the process of atomization, and therefore the effectiveness of creating free atoms in the determined element. This effect is extremely important with consideration to the entire chemical measurement and must be taken into account when validating a given measurement procedure.

The homogeneity of a sample (heterogeneity of composition) significantly influences the determination of the representative portion of the examined material. While planning analytical conduct, an analyst must allow for the heterogeneity of a sample's composition; hence, for solid samples, an analytical sample should be sufficiently large so that grain size is not a source of heterogeneity.

The stability of a sample determines the measurement duration. In some cases, the composition of a sample can change even over several minutes, hence the necessity for exact knowledge concerning how the sample behaves over time.

The preparation of a sample is the most important element causing difficulty in maintaining traceability. Every physicochemical operation disrupts the chain of traceability, which implies a necessity for a detailed plan of action.

Correct realization of a measurement depends primarily on the efficiency of the measuring instrument used and the maintenance of suitable measurement conditions. For example, a pH measurement requires calibration of an instrument and measurements at a suitable temperature.

Determination of the uncertainty of a result is an integral part of traceability assurance. Uncertainties in reference standards comprise the uncertainty of a result obtained by a comparison with these standards.

Traceability should be shown for each parameter of a given procedure and should be carried out by calibration with suitable standards.

A procedure enabling the determination of correlations between the value of a signal (indicated by an instrument) and the concentration of the examined substance in the sample is called calibration.

It is necessary to use reference standards for which traceability may be shown and which have known uncertainty.

An important part at this step is played by reference materials, which can assure traceability to standards, that is to say, which obtain traceability, and consequently international agreement on measurements [9].

In practice, traceability for chemical measurements can be determined in two ways [10]:

- By comparing an obtained value with reference measurements.
- By referring an obtained value to reference standards, which in turn have a connection with the value obtained in reference measurements.

Reference values should come from expert laboratories with good international reputations.

For trace analysis, fulfilling the traceability requirement for a typical analytical procedure demands the use of a matrix reference materials. The traceability of measurement results depends on, among other things, the proper functioning of instruments, which can be assured by calibration using suitable calibrants. The calibration step is used for [11,12]

- Assuring the correct performance of an instrument (instrument calibration)
- Determining a clear dependence between a determined signal and a determined property (analytical calibration)

For reference materials reproducing the chemical properties, the problem of traceability assurance involves the accessibility of standards with a required level of analyte concentration, determined with a suitable accuracy (higher than the accuracy in the applied analytic methodology).

Here it must be remembered that very frequently, analytes in samples occur in trace or ultratrace amounts, and preparing suitable reference materials poses an immense challenge.

This has an undoubtable influence on the cost of preparing reference materials.

Many reference materials may have properties that, for various reasons, cannot be measured in units of mass or quantity, nor determined by means of precisely defined physical and chemical measurement methods. Examples of such reference materials are biological reference materials attributed to a respective international unit by the World Health Organization, and also technological reference materials [13].

Assuring traceability, and hence assuring the reliability of measurements, is an element of analytical chemistry which is currently given considerable attention. That is why the notion of traceability and the associated notion of uncertainty are also two key problems in present-day metrology in analytical chemistry. For the purpose of obtaining a full and correct picture, traceability should be considered in four ways [14]:

- The traceability of analytical results, that is, the assurance of the obtained analytical results referred to specific reference materials by an uninterrupted chain of comparisons of uncertainties associated with suitable reference materials (certification and history of their production).
- The traceability of the applied standards, that is, the properties of standard values that can be related to reference materials by an uninterrupted chain of comparisons of uncertainties associated with suitable reference materials and supplied documentary evidence giving the history of their production (in which significant properties such as homogeneity, stability, and origin must be clearly presented).
- The traceability of an instrument, that is, a detailed and up-to-date history of the instruments, containing descriptions of their installation, damage, number of hours used, sample processing, and other parameters associated

with the specific instrument, with special attention paid to maintenance, calibration, and repairs.

- The traceability of analytical methodology (procedures), that is, the possibility of obtaining traceable results after a correct process of validating all analytical conduct.

In compliance with the requirements stated in the EURACHEM/CITAC Guide [15], in order to determine the traceability of a given analytical procedure, it is necessary to

- Determine the measurand
- Select a suitable measurement procedure and record a respective model equation
- Prove (by validation) the correctness of the selected measuring conditions and the model equation
- Determine a strategy for proving traceability by selecting suitable standards and determining procedure calibration
- Determine the uncertainty of the applied measurement procedure

As noted earlier, one of the most important tools used for the purpose of traceability assurance in chemical measurements is the use of certified reference materials, which are extremely useful for

- Estimating the accuracy of new analytical procedures
- Comparing different methods
- Comparing and testing the competence of different laboratories

Realization of traceability in chemical measurements by means of reference materials can be carried out using pure standard substances for calibration or suitable certified reference materials. It is very important to purchase reference material from a reputable distributor, which will assure the maintenance of traceability for a given value together with a given uncertainty value. The most important criteria for the selection of reference materials are primarily the agreement of matrix and concentrations of the determined substance. Moreover, it is necessary to allow for uncertainty provided by the manufacturer and to estimate to what extent this will be important in the uncertainty budget of the applied measurement procedure.

4.4 CONCLUSION

The main sense of traceability is to enable comparability of measurement results—either compare results of the measurements on the same sample or compare results on different samples [12]. In theory, all measurements can be tracked back to the base seven SI units [16]. Traceability is highly connected with uncertainty, comparability, utility, reliability, and validity.

Traceability and uncertainty are necessary parameters for obtaining reliable results.

REFERENCES

1. International vocabulary of metrology—Basic and general concepts and associated terms (VIM), Joint Committee for Guides in Metrology, JCGM 200, 2008.
2. Valcárcel M., and Rios A., Traceability in chemical measurements for the end users, *Trends Anal. Chem.*, 18, 570–576, 1999.
3. Quality Management and Quality Assurance Vocabulary, ISO 8402, Geneva, 1994.
4. Walsh M.C., Moving from official to traceable methods, *Trends Anal. Chem.*, 18, 616–623, 1999.
5. De Bièvre P., Kaarls R., Peiser H.S., Rasberry S.D., and Reed W.P., Protocols for traceability in chemical analysis. Part II: Design and use, *Accred. Qual. Assur.*, 2, 270–274, 1997.
6. De Bièvre P., and Taylor P.D.P., Traceability to the SI of amount-of substance measurements: From ignoring to realizing a chemist's view, *Metrologia*, 34, 67–75, 1997.
7. Thomson M., Comparability and traceability in analytical measurements and reference materials, *Analyst*, 122, 1201–1205, 1997.
8. King B., The practical realization of the traceability of chemical measurements standards, *Accred. Qual. Assur.*, 5, 429–436, 2000.
9. ISO guide 35: Certification of reference materials. General and statistical principles. ISO, Geneva, 1989.
10. Bulska E., and Taylor Ph., On the importance of metrology in chemistry, in Namieśnik J., Chrzanowski W., and Żmijewska P. (Eds), *New Horizons and Challenges in Environmental Analysis and Monitoring*, CEEAM, Gdańsk, 2003.
11. Valcárcel M., and Rios A., Is traceability an exclusive property of analytical results? An extended approach to traceability in chemical analysis, *Fresenius J. Anal. Chem.*, 359, 473–475, 1997.
12. Williams A., Traceability and uncertainty: A comparison of their application in chemical and physical measurement, *Accred. Qual. Assur.*, 6, 73–75, 2001.
13. ISO Guide 30, Trends and definitions used in connections with reference materials, ISO, Geneva, 1992.
14. Marschal A., Andrieux T., Compagon P.A., and Fabre H., Chemical metrology: QUID? *Accred. Qual. Assur.*, 7, 42–49, 2002.
15. EURACHEM/CITAC Guide, Traceability in Chemical Measurements, A guide to achieving comparable results in chemical measurement, 2003.
16. Buzoianu M., and Aboul-Enein H.Y., The traceability of analytical measurements, *Accred. Qual. Assur.*, 2, 11–17, 1997.

5 Uncertainty

5.1 DEFINITIONS [1–3]

Combined standard uncertainty $u_{c(y)}$: Standard measurement uncertainty that is obtained using the individual standard measurement uncertainties associated with the input quantities in a measurement model; standard uncertainty of a result y of a measurement when the result is obtained from the values of many other quantities equal to the positive square root of a sum of terms, the terms being the variances or covariances of these other quantities weighted according to how the measurement result varies with these quantities.

Coverage factor k: Number larger than 1 by which a combined standard measurement uncertainty is multiplied to obtain an expanded measurement uncertainty. A coverage factor is typically in the range of 2–3, and for an approximately 95 percent level of confidence, $k = 2$.

Definitional uncertainty: Component of measurement uncertainty resulting from the finite amount of detail in the definition of a measurand.

Expanded uncertainty U: Product of a combined standard measurement uncertainty and a factor larger than the number 1.

Relative uncertainty $u_{r(x_i)}$: Standard measurement uncertainty divided by the absolute value of the measured quantity value.

Standard uncertainty $u_{(x_i)}$: Uncertainty of a result x_i of a measurement expressed as a standard deviation.

Type A evaluation (of uncertainty): Evaluation of a component of measurement uncertainty by a statistical analysis of measured quantity values obtained under defined measurement conditions.

Type B evaluation (of uncertainty): Evaluation of a component of measurement uncertainty determined by means other than a Type A evaluation of measurement uncertainty.

Uncertainty budget: Statement of a measurement uncertainty, of the components of that measurement uncertainty, and of their calculation and combination.

Uncertainty of measurement: Nonnegative parameter characterizing the dispersion of the quantity values being attributed to a measurand, based on the information used.

5.2   INTRODUCTION

Decisions made in many fields of science and other domains of life are based on the results of analytical studies. It is therefore obvious that their quality is increasingly important.

Uncertainty of measurement is a component of uncertainty for all individual steps of an analytical procedure [4–7]. Hence, it is necessary to determine the sources and types of uncertainty for all these steps [8–10].

The main sources of uncertainty during sample analysis while using an appropriate analytical procedure may be [11]

- Inaccurate or imprecise definition of the measurand
- Lack of representativeness at the step of collecting a sample from an examined material object
- Inappropriate methodology of determinations
- Personal deviations in reading the analog signals
- Not recognizing the influence of all the external factors on the result of an analytical measurement
- Uncertainty associated with the calibration of an applied measurement instrument
- Insufficient resolution of the applied measurement instrument
- Uncertainties associated with the applied standards or reference materials
- Uncertainties of parameters determined in separate measurements and which are used in calculating the final result; such as physicochemical constants
- Approximations and assumptions associated with using a given instrument, applied during measurement
- Fluctuations of the measurement instrument gauge, over the course of repeated measurements, with seemingly identical external conditions

There is a difference between measurement error and uncertainty. The error is a difference between the determined and expected values, and uncertainty is a range into which the expected value may fall within a certain probability. So the uncertainty cannot be used to correct a measurement result.

This difference is schematically presented in Figure 5.1 [11].

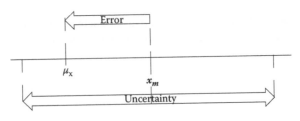

FIGURE 5.1   Schematic presentation of difference between error and measurement uncertainty. (From Konieczka, P., *Crit. Rev. Anal. Chem.*, 37, 173–190, 2007.)

5.3 METHODS OF ESTIMATING MEASUREMENT UNCERTAINTY

There are several approaches for uncertainty estimation [12,13]:

- Bottom-up: Based on an identification, quantification, and combination of all individual sources of uncertainty of measurement. The overall uncertainty is derived from the uncertainties of individual components. This method has high complexity and because of that it needs considerable time and effort; this approach is adapted by EURACHEM [2,14].
- Fitness-for-purpose: Based on a definition of single parameter called the fitness function, which has the form of algebraic expression, and describes the relation between uncertainty and analyte content. Calculation of uncertainty for the result of measurement is very easy and less time-consuming than a *bottom-up* approach.
- Top-down: Based on data obtained from interlaboratory studies (precision).
- Validation-based: Based on inter- or intralaboratory validation studies (precision, trueness, robustness).
- Robustness-based: Based on robustness tests from interlaboratory studies.

5.3.1 PROCEDURE FOR ESTIMATING THE MEASUREMENT UNCERTAINTY ACCORDING TO THE GUIDE TO THE EXPRESSION OF UNCERTAINTY IN MEASUREMENT

Determining the uncertainty of a measurement increases its reliability, and in turn allows comparison of results obtained in interlaboratory studies and helps users to decide the significance of any difference between the obtained result and the reference value.

According to the Guide to the Expression of Uncertainty in Measurement [2], in order to determine the uncertainty of analysis result, the following conditions must be satisfied:

- a. The measurement procedure and the measurand must be defined. The measurand in a given measurement must be clearly defined, along with the unit in which it is expressed. The observed quantity and the searched parameter (result) must also be clearly described.
- b. Modeling (usually mathematical modeling) must be applied to calculate the analysis result based on the measured parameters.
  An appropriate mathematical model ties the value of a determination result (the one to be determined) with the observed values (measurement values). The relation is described as follows:

$$y = f(x_1, x_2 \cdots x_n) \tag{5.1}$$

where
  y is the value of a result
  $x_1, x_2...x_n$: measurement values

c. Values must be assigned to all the possible parameters that could affect the final result of the analysis, and the standard uncertainty for each of them must be determined.

Each measurand has a name, unit, value, standard uncertainty, and number of degrees of freedom. As noted before, there are two methods for calculating standard uncertainty. Type A uncertainty is equal to a standard deviation of an arithmetic mean. Type B uncertainty is strictly associated with the probability distribution described by the distribution of a variable.

For example, when a variable has a rectangular distribution, such as in the case of a standard's purity, the variable may assume (with equal probability) a value in the range $\langle -a, +a \rangle$, and the calculated standard uncertainty is $\dfrac{a}{\sqrt{3}}$ (where a is the midpoint of the range $\langle -a, +a \rangle$).

When a variable has a triangular distribution, which means that the value is in the range $\langle -a, +a \rangle$, but the occurrence of the mean value from the range is the most probable, the calculated standard uncertainty is $\dfrac{a}{\sqrt{6}}$.

Example 5.1

Problem: Calculate standard uncertainty for the concentration of magnesium in a standard solution, based on data given by producer.
Data: Standard solution concentration $C_{st} = 1001 \pm 2$ mg/dm³.
Solution: Due to no additional information, we assume a uniform distribution.

$$u_{(C_{st})} = \frac{2}{\sqrt{3}}$$

$$u_{(C_{st})} = 1.2 \text{ mg/dm}^3$$

Excel file: exampl_uncert01.xls

Example 5.2

Problem: Calculate the standard uncertainty for the concentration of magnesium in a standard solution, based on data given by the producer. Uncertainty given by the manufacturer has been calculated for coverage factor $k = 2$.
Data: Standard solution concentration $C_{st} = 1001 \pm 2$ mg/dm³.
Solution: Because value for k is given, the standard uncertainty is calculated accordingly:

$$u_{(C_{st})} = \frac{2}{k}$$

$$u_{(C_{st})} = 1 \text{ mg/dm}^3$$

Excel file: exampl_uncert02.xls

Example 5.3

Problem: Calculate the standard uncertainty for the determination of volume 500 cm³ using a volumetric flask, based on data given by the manufacturer. Assume triangular distribution.

Data: Volume V_{fl} = 500 ± 0.8 cm³.

Solution:

$$u_{(V_{fl})} = \frac{0.8}{\sqrt{6}}$$

$$u_{(V_{fl})} = 0.32 \text{ cm}^3$$

Excel file: exampl_uncert03.xls

d. The applied principles of uncertainty propagation in calculating the standard uncertainty of an analytical result.

For a given mathematical model that binds the final results of analysis with measured parameters (Equation 5.1), standard uncertainty is calculated by using the principle of uncertainty propagation expressed in the following formula:

$$u_{c(y)}^2 = \sum_{i=1}^{n} \left(\frac{\delta f}{\delta x_i} \right)^2 u_{(x_i)}^2 \tag{5.2}$$

When the value of an analytical result is the sum or difference of the measurement values,

$$y = x_1 + x_2 + \ldots + x_n \tag{5.3}$$

then the value of the combined uncertainty is described by the following equation:

$$u_{c(y)} = \sqrt{u_{(x_1)}^2 + u_{(x_2)}^2 + \ldots + u_{(x_n)}^2} \tag{5.4}$$

Due to the very frequent occurrence of individual measurement values being expressed in different units, it is more convenient to apply relative uncertainties. A relative uncertainty is described by the following relation:

$$u_{r(x_i)} = \frac{u_{(x_i)}}{x_i} \tag{5.5}$$

If the value of the analytical result is a quotient/product of the measurement values,

$$y = \frac{x_1 \times x_2 \times \ldots}{x_3 \times \ldots} \qquad (5.6)$$

then the value of the combined relative uncertainty is described by the following equation:

$$u_{r(y)} = \sqrt{u_{r(x_1)}^2 + u_{r(x_1)}^2 + \ldots + u_{r(x_n)}^2} \qquad (5.7)$$

e. Presentations of the final result of the analysis are: *result ± expanded uncertainty* (after using an appropriate *k* factor).
   Uncertainty calculated according to the aforementioned equation is a combined standard uncertainty of the final determination. To calculate the value of the expanded uncertainty, the combined standard uncertainty should be multiplied by an appropriate coverage factor *k*.
   Therefore, the final result of an analysis comprises the following:
   • Determination of the measured value and its unit
   • The result with the expanded uncertainty value ($y \pm U$, along with units for *y* and *U*)
   • *k* factor value, for which the expanded uncertainty has been calculated

Thus, a correctly presented result of an analysis should be as follows:

$$c_{NaOH} \pm U(k=2) = 0.1038 \pm 0.0017 \ [\text{mol/dm}^3]$$

or

$$c_{NaOH} \pm U(k=2) = 0.1038 \ \text{mol/dm}^3 \pm 1.6[\%]$$

Example 5.4

Problem: A standard Mg^{2+} solution was prepared with a basic diluted solution with a concentration of 1001 ± 2 mg/dm³. With the aim of obtaining a standard solution with a concentration of around 0.5 mg/dm³, the basic solution was diluted as follows:

   We took 1 cm³ of basic sample solution by using a pipette with a volume of 1 cm³ and transferred it to a volumetric flask with a volume of 100 cm³. After filling the flask to the line and mixing the solution, 5 cm³ of solution was taken from it with the help of a pipette with a volume of 5 cm³ and was transferred to a volumetric flask with a volume of 100 cm³ and after being filled to the line, a standard solution was obtained with the predetermined concentration.

To establish a uniform distribution for each of the measured parameters, calculate the following:
- The value of the combined and expanded uncertainty (for $k = 2$) for the obtained standard solution concentration, and present indication results.
- The participation percentage of each of the standard uncertainty values in the determined values of the combined uncertainty.

Data:

			Unit
Standard solution concentration	C_{st}	1001	mg/dm³
Pipette 1 volume	V_{p1}	1	cm³
Flask 1 volume	V_{f1}	100	cm³
Pipette 2 volume	V_{p2}	5	cm³
Flask 2 volume	V_{f2}	100	cm³
Uncertainty of single measurement	$u(C_{st})$	2	mg/dm³
	$u(V_{p1})$	0.02	cm³
	$u(V_{f1})$	0.2	cm³
	$u(V_{p2})$	0.03	cm³
	$u(V_{f2})$	0.2	cm³
Distribution	rectangular (R) or triangular (T)		R

Solution:

x_i	u_r	Relative Uncertainty Contribution*,%
C_{st}	0.0012	0.89%
V_{p1}	0.012	89.29%
V_{f1}	0.0012	0.89%
V_{p2}	0.0035	8.04%
V_{f2}	0.0012	0.89%
k	2	

*Calculated as: $\dfrac{\left(u_r(x_i)\right)^2}{\left(u_r(c)\right)^2}$ [%]

		Unit
c	0.5005	mg/dm³
$u_r(c)$	0.012	
$U(c)(k = 2)$	0.012	mg/dm³
$U\%$	2.4%	
Result	0.501 ± 0.012	mg/dm³

Excel file: exampl_uncert04.xls

Example 5.5

Problem: A standard Mg^{2+} solution was prepared with a basic diluted solution with a concentration of 1001 ± 2 mg/dm³. With the aim of obtaining a standard solution with a concentration of around 0.5 mg/dm³, the basic solution was diluted as follows:

A basic standard solution of 10 cm³ was taken using a pipette with a volume of 10 cm³ and was transferred to a volumetric flask with a volume of 100 cm³. After filling the flask to the line and mixing the solution, 5 cm³ of solution was taken from it with the help of a pipette with a volume of 5 cm³ and was transferred to a volumetric flask with a volume of 100 cm³. After filling the flask to the line and mixing the solution, 10 cm³ of solution was taken from it with the help of a pipette with a volume of 10 cm³ and was transferred to a volumetric flask with a volume of 100 cm³ and after being filled to the line a standard solution was obtained with the predetermined concentration.

To establish a uniform distribution for each of the measured parameters, calculate the following:
- The value of the combined and expanded uncertainty (for $k = 2$) for the obtained standard solution concentration, and present indication results.
- The participation percentage of each of the standard uncertainty values in the determined values of the combined uncertainty.

Data:

			Unit
Standard solution concentration	C_{st}	1001	mg/dm³
Pipette 1 volume	V_{p1}	10	cm³
Flask 1 volume	V_{f1}	100	cm³
Pipette 2 volume	V_{p2}	5	cm³
Flask 2 volume	V_{f2}	100	cm³
Pipette 3 volume	V_{p2}	10	cm³
Flask 3 volume	V_{f2}	100	cm³
Uncertainty of single measurement	$u(C_{st})$	2	mg/dm³
	$u(V_{p1})$	0.04	cm³
	$u(V_{f1})$	0.2	cm³
	$u(V_{p2})$	0.03	cm³
	$u(V_{f2})$	0.2	cm³
	$u(V_{p2})$	0.04	cm³
	$u(V_{f2})$	0.2	cm³
Distribution	rectangular (R) or triangular (T)		R

Solution:

x_i	u_r	Relative Uncertainty Contribution, %
C_{st}	0.0012	4.75%
V_{p1}	0.0023	19.05%
V_{f1}	0.0012	4.76%
V_{p2}	0.0035	42.86%
V_{f2}	0.0012	4.76%
V_{p3}	0.0023	19.05%
V_{f3}	0.0012	4.76%
k	2	

		Unit
c	0.5005	mg/dm³
$u_r(c)$	0.0053	
$U(c)$	0.0053	mg/dm³
$U\%$	1.1%	
Result	0.5005 ± 0.0053	mg/dm³

Excel file: exampl_uncert05.xls

Example 5.6

Problem: A standard sample was weighed for the preparation of a standard solution. The mass measurement was carried out using an analytical scale, for which its producer gave a measurement uncertainty of 0.5 mg.

The mass was calculated as the difference of two mass measurements: Gross (container with a sample = 332.55 mg) and net (container = 187.72 mg). Calculate the standard uncertainty of the mass measurement. Assume a rectangular distribution of the parameter. Assuming the value of the coverage factor to be 2, calculate the expanded uncertainty of the mass measurement. Give a correct presentation of the mass measurement result.

Data:

			Unit
Mass (tarra)	m_{tarra}	187.72	mg
Mass (brutto)	m_{brutto}	332.55	mg
Uncertainty of single measurement	$u(m_{tarra})$	0.5	mg
	$u(m_{brutto})$	0.5	mg
Distribution	rectangular (R) or triangular (T)		R

Solution:

m_{netto}	144.83	mg
$u(m_{netto})$	0.41	mg
k	2	
$U(m_{netto})$	0.82	mg
Result	144.83 ± 0.82	mg

Excel file: exampl_uncert06.xls

Example 5.7

Problem: The weighed standard sample (Example 5.6) was put into a measurement flask (250 cm³) for which the manufacturer provided an uncertainty value equal to 0.4 cm³. Calculate the combined standard uncertainty of the obtained standard solution concentration. Assume a rectangular distribution of the parameters. Assuming the value of the coverage factor to be 2, calculate the expanded uncertainty of the concentration. Give a correct presentation of the result.

Data:

			Unit
Mass (tarra)	m_{tarra}	187.72	mg
Mass (brutto)	m_{brutto}	332.55	mg
Flask volume	V_{flask}	250	cm³
Uncertainty of single	$u(m_{tarra})$	0.5	mg
measurement	$u(m_{brutto})$	0.5	mg
	$u(V_{flask})$	0.4	cm³
Distribution	rectangular (R) or triangular (T)		R

Solution:

Concentration	0.5793	mg/cm³
u_r(concentration)	0.0020	
k	2	
U(concentration)	0.0023	mg/cm³
Result	0.5793 ± 0.0023	mg/cm³

Excel file: exampl_uncert07.xls

Example 5.8

Problem: The obtained standard solution (Example 5.7) was dissolved by a 1:10 ratio, sampling 1 cm³ of the original solution using a pipette for which the manufacturer provided an uncertainty value of 0.2 cm³, and dissolving in a measurement flask (10 cm³) for which the manufacturer provided an uncertainty value 0.05 cm³.

Calculate the combined standard uncertainty of the obtained standard solution concentration. Assume a rectangular distribution of parameters. Assuming the value of the coverage factor to be 2, calculate the expanded uncertainty of the concentration. Give a correct presentation of the result.

Data:

			Unit
Mass (tarra)	m_{tarra}	187.72	mg
Mass (brutto)	m_{brutto}	332.55	mg
Flask volume	V_{flask1}	250	cm³
Flask volume	V_{flask2}	10	cm³
Pipette	$V_{pipette}$	1	cm³
Uncertainty of single	$u(m_{tarra})$	0.5	mg
measurement	$u(m_{brutto})$	0.5	mg
	$u(V_{flask1})$	0.4	cm³
	$u(V_{flask2})$	0.05	cm³
	$u(V_{pipette})$	0.2	cm³
Distribution	rectangular (R) or triangular (T)		R

Solution:

Concentration	0.0579	mg/cm³
u_r(concentration)	0.12	
k	2	
U(concentration)	0.13	mg/cm³
Result	0.058 ± 0.013	mg/cm³

Excel file: exampl_uncert08.xls

5.4 TOOLS USED FOR UNCERTAINTY ESTIMATION

Correct estimation of uncertainty needs an understanding of whole analytical procedure by analyst. The most helpful tools used for that are [4,6] the following:

- Flow diagram, which are drawn on the basis of information presented in detail in a standard operating procedure.
- Ishikawa, or cause-and-effect, or fishbone diagram, which show the influence parameters (sources of uncertainty) of a whole analytical procedure [15,16].

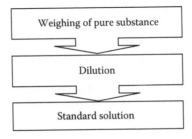

FIGURE 5.2 Flow diagram for the procedure of preparation of standard solution.

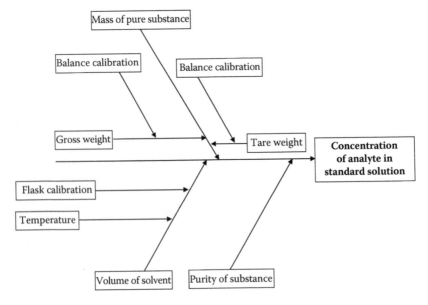

FIGURE 5.3 Ishikawa diagram for the procedure of preparation of standard solution.

The flow diagram and the Ishikawa diagram for the procedure of preparation of a standard solution are presented in Figures 5.2 and 5.3, respectively.

5.5 UNCERTAINTY AND CONFIDENCE INTERVAL

In some cases, the value of uncertainty can be estimated as a confidence interval. The basic principle of the uncertainty propagation is underlining the influence of the quantity with the highest value.

Therefore, if one of the parameters has a dominating influence over the uncertainty budget, calculation of uncertainty may be limited to the calculation based on the value of that parameter. If that dominating parameter is the repeatability of measurements, then the expanded uncertainty may be calculated according to the following relation:

$$U = k\frac{SD}{\sqrt{n}} \tag{5.8}$$

where
  SD: Standard deviation
  n: Number of measurements

On the other hand, the value of confidence interval could be calculated as

$$\Delta x_{s'r} = t(\alpha, f)\frac{SD}{\sqrt{n}} \tag{5.9}$$

For a level of significance of $\alpha = 0.05$, the coverage factor $k = 2$
For a level of significance of $\alpha = 0.05$ and the number of degrees of freedom $f \to \infty$,
the parameter $t \approx 2$—Table A.1 (in the appendix)
Given these conditions, the aforementioned equations are thus consistent

Example 5.9

Problem: The concentration of mercury was determined in water using the cold vapor atomic absorption spectrometry (CVAAS) technique. The series involved four determinations. Considering the unrepeatability as the main component of the uncertainty budget, calculate the expanded uncertainty of the determination result for $k = 2$. Provide a correct presentation of the determination result.
Data: Result series, $\mu g/dm^3$:

	Data
1	71.53
2	72.14
3	77.13
4	76.54

Solution:

Mean	74.335	$\mu g/dm^3$
SD	2.91	$\mu g/dm^3$
k	2	
U	2.9	$\mu g/dm^3$
Result	74.3 ± 2.9	$\mu g/dm^3$

Excel file: exampl_uncert09.xls

5.6   CALIBRATION UNCERTAINTY

A decisive majority of analytical measurements involve a calibration step, which is associated with the relative (comparative) character of measurements. At the calibration step, a calibration curve technique is usually used, which is determined using linear regression. This step of the analytical procedure has influence on the combined uncertainty of the determination result for the real sample. Standard uncertainty due to that step of the analytical procedure should be included in the uncertainty budget.

There are four sources of uncertainty due to the calibration step which can influence the standard uncertainty of a single measurement $u_{(x_{smpl})}$ [8,17–19]:

- Repeatability of reading the value of a signal y both for standard samples (based on measurements for which the calibration curve is determined) and for study samples— $u_{(x_{smpl},y)}$
- Uncertainty due to the determination of the reference value for standard samples— $u_{(x_{smpl},x_{std_i})}$
- The influence of the manner of preparing the standard samples, usually using a method of consecutive dilutions
- Incorrect approximation of measurement points using a regression curve

Figure 5.4 presents an example of a calibration graph along with the marked uncertainty values associated with both the reading of the signal values and the reference values.

Using a calibration curve drawn based on Equations 1.63–1.68 in Chapter 1, it is possible to determine and identify the uncertainty of the determined regression

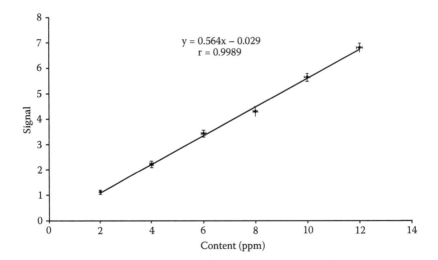

FIGURE 5.4   An example of a calibration graph along with the marked uncertainty values associated both with the reading of the signal values and the reference values.

curve through the determination of confidence intervals. Those intervals are determined using a correlation that is described by the following equation:

$$\Delta y_i = Y \pm SD_{xy} \cdot t_{(\alpha, f=n-2)} \sqrt{\frac{1}{n} + \frac{(x_i - x_m)^2}{Q_{xx}}} \tag{5.10}$$

where

Δy_i: The confidence interval of the calculated value Y for a given value x_i

Y-values: Calculated based on the regression curve equation for given values x_i

SD_{xy}: Residual standard deviation

$t_{(\alpha, f=n-2)}$: Student's t test parameter

n: The total number of standard samples used for the determination of the calibration curve (number of points)

x_i: Value x for Δy_i is calculated

x_m: Mean value x (x is most frequently the analyte concentration and is the mean of all the concentrations of a standard solution for which the measurement was made in order to make a standard curve)

Q_{xx}: Parameter calculated according to a relation described by the equation

$$Q_{xx} = \sum_{i=1}^{n} (x_i - x_m)^2 \tag{5.11}$$

Standard uncertainty for x_{smpl} due to the uncertainty of calibration and linear regression method $u(x_{smpl}, y)$ may be calculated using the determined regression parameters according to the following relationship:

$$u_{(x_{smpl}, y)} = \frac{SD_{xy}}{b} \sqrt{\frac{1}{p} + \frac{1}{n} + \frac{(x_{smpl} - x_m)^2}{Q_{xx}}} \tag{5.12}$$

where

$u_{(x_{smpl}, y)}$: Standard uncertainty for the determination of the x_{smpl} concentration due to the application of the determined calibration correlation

b: The slope of the calibration curve

p: The number of measurements (repetitions) carried out for a given sample

Figure 5.5 presents a calibration curve along with the marked confidence intervals and the determined uncertainty value for the determination of an analyte's concentration in an examined sample.

The value of uncertainty for the determination of analyte concentration in the applied standard samples is usually significantly smaller compared to the uncertainty

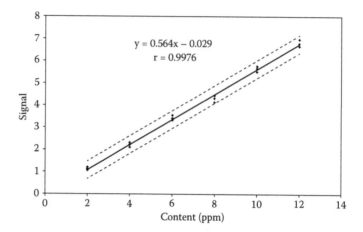

FIGURE 5.5 A calibration curve along with the marked confidence intervals and the determined uncertainty value for the determination of an analyte's concentration in an examined sample.

associated with the calculation of analyte concentration based on the following determined calibration function:

$$u_{(x_{smpl}x_{std_i})} \ll u_{(x_{smpl},y)} \tag{5.13}$$

Therefore, its value may be estimated by only considering the number of standard samples used at the calibration stage. Because usually only one basic standard is used and then appropriate standard solutions are made (consecutive dilutions), standard uncertainty due to the application of standard solutions at the calibration step may be described by the following equation:

$$u_{(x_{smpl},x_{std_i})} \approx \frac{u_{(x_{std_i})}}{n} \tag{5.14}$$

Such an uncertainty value does not allow for the uncertainty associated with the manner of standard sample preparation. If each standard sample is prepared by consecutive dilutions, then the uncertainty budget must allow for the standard uncertainties associated with the step of standard sample preparation. Usually, the standard uncertainty of a result, associated with an applied calibration technique, requires only the value $u_{(x_{smpl},y)}$.

Example 5.10

Problem: A calibration curve was determined using determinations of analyte concentration in samples of six standard solutions, making three independent measurements for each of the solutions.

Calculate:

- Regression parameters of the calibration curve
- Confidence intervals
- Uncertainty value of the determination value for the real sample due to calibration, for which three independent measurements were made and the result was calculated using the determined curve

Data: Results, ppm:

		Data	
		x, ppm	y, signal
1		2	1.12
2		2	1.2
3		2	1.08
4		4	2.11
5		4	2.32
6		4	2.23
7		6	3.33
8		6	3.54
9		6	3.41
10		8	4.12
11		8	4.32
12		8	4.44
13		10	5.67
14		10	5.76
15		10	5.51
16		12	6.97
17		12	6.78
18		12	6.66

Result for sample	7.59	ppm
Number of measurements for sample	3	

Solution:

n	18
Slope: b	0.564
Intercept: a	−0.029
Residual standard deviation: $SD_{x.y}$	0.143
Regression coefficient: r	0.9976
Q_{xx}	210
Uncertainty for result due to calibration	0.16 ppm
Relative uncertainty for result due to calibration	2.1%
$t(\alpha = 0.05; f = n - 2)$—from Table A.1 (in the appendix)	2.12

Graph:

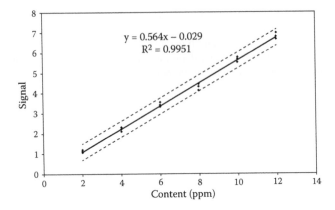

$$y = 0.564x - 0.029$$
$$R^2 = 0.9951$$

(Y-axis: Signal; X-axis: Content (ppm))

Excel file: exampl_uncert10.xls

5.7 CONCLUSION

Each analytical result derives from a conducted measurement. The ultimate goal for an analyst is to obtain a result that will most reliably reflect the expected (actual, real) value. The certainty of the analytical result depends on the uncertainties occurring at all the steps of an analytical procedure, the basic tool for any analyst.

The most crucial parameter affecting a measurement result's uncertainty is the parameter with the highest uncertainty value. Therefore, it is necessary to determine the sources and types of uncertainty for individual steps of an analytical procedure, and more exactly for each measurand. Combined uncertainty covers all sources of uncertainty that are relevant for all analyte concentration levels. It is a "key indicator" of both fitness-for-purpose and reliability of results.

Uncertainty is a basic property of each measurement. Uncertainty occurs always and at any step of a measurement procedure. Hence, it is not a property that should result in additional difficulties during the measurement procedure.

REFERENCES

1. International vocabulary of metrology—Basic and general concepts and associated terms (VIM), Joint Committee for Guides in Metrology, JCGM 200, 2008.
2. Guide to the Expression of Uncertainty in Measurement (GUM), ISO, Geneva, 1993.
3. Williams A., Introduction to measurement uncertainty in chemical analysis, *Accred. Qual. Assur.*, 3, 92–94, 1998.
4. Populaire A., and Campos Gimenez E., A simplified approach to the estimation of analytical measurement uncertainty, *Accred. Qual. Assur.*, 10, 485–493, 2005.
5. Roy S., and Fouillac A.-M., Uncertainties related to sampling and their impact on the chemical analysis of groundwater, *Trends Anal. Chem.*, 23, 185–193, 2004.
6. Meyer V.R., Measurement uncertainty, *J. Chromatogr. A*, 1158, 15–24, 2007.
7. Kadis R., Evaluating uncertainty in analytical measurements: The pursuit correctness, *Accred. Qual. Assur.*, 3, 237–241, 1998.

8. Conti M.E., Muse O.J., and Mecozzi M., Uncertainty in environmental analysis: Theory and laboratory studies, *Int. J. Risk. Assess. Manag.*, 5, 311–335, 2005.

9. Love J.L., Chemical metrology, chemistry and the uncertainty of chemical measurements, *Accred. Qual. Assur.*, 7, 95–100, 2002.

10. Armishaw P., Estimating measurement uncertainty in an afternoon. A case study in the practical application of measurement uncertainty, *Accred. Qual. Assur.*, 8, 218–224, 2003.

11. Konieczka P., The role of and place of method validation in the quality assurance and quality control (QA/QC) System, *Crit. Rev. Anal. Chem.*, 37, 173–190, 2007.

12. Sahuquillo A., and Rauret G., Uncertainty and traceability: The view of the analytical chemist, in Fajgelj A., Belli M., and Sansone U. (eds.), *Combining and Reporting Analytical Results*, RSC, Springer, Berlin, 2007.

13. Taverniers I., Van Bockstaele E., and De Loose M., Trends in quality in the analytical laboratory. I. Traceability and measurement uncertainty of analytical results, *Trends Anal. Chem.*, 23, 480–490, 2004.

14. Ellison S.L.R., Rosslein M., and Williams A., EURACHEM, Quantifying Uncertainty in Analytical Measurements, 2nd ed., 2000.

15. Ellison S.L.R., and Barwick V.J., Using validation data for ISO measurements uncertainty estimation. Part 1. Principles of an approach using cause and effect analysis, *Analyst*, 123, 1387–1392, 1998.

16. Kufelnicki A., Lis S., and Meinrath G., Application of cause-and-effect analysis to potentiometric titration, *Anal. Bioanal. Chem.*, 382, 1652–1661, 2005.

17. Danzer K., and Currie L.A., Guidelines for calibration in analytical chemistry, *Pure Appl. Chem.*, 70, 993–1014, 1998.

18. Bonate P.L., Concepts in Calibration Theory, Part IV: Prediction and Confidence Intervals, *LC-GC*, 10, 531–532, 1992.

19. Miller J.N., Basic Statistical Methods for Analytical Chemistry Part 2. Calibration and Regression Methods, *Analyst.*, 116, 3–14, 1991.

6 Reference Materials

6.1 DEFINITIONS [1,2]

Certified Reference Material (CRM): Reference material, accompanied by documentation issued by an authoritative body and providing one or more specified property values with associated uncertainties and traceabilities, using valid procedures.

Homogeneity: Condition of having a uniform structure or composition with respect to one or more specified properties. RM is said to be homogeneous with respect to a specified property if the property value, as determined by tests on samples of specified size, is found to lie within the specified uncertainty limits, the samples being taken either from different supply units (bottles, packages, etc.)—between-bottle homogeneity, or from a single supply unit—within-bottle homogeneity.

Reference material (RM): Material, sufficiently homogeneous and stable with reference to specified properties, that has been established to be fit for its intended use in measurement or in examination of nominal properties.

Stability: Ability of a reference material, when stored under specified conditions, to maintain a stated property value within specified limits for a specified period of time.

6.2 INTRODUCTION

Reference materials play a significant role in all the elements of the quality assurance system that evaluates the reliability of measurement results. The range of their application varies [3–7]:

- Validation of analytical procedures, where reference materials are used to determine precision and accuracy
- Interlaboratory comparisons, where they are applied as subject matter for studies
- Estimating the uncertainty of a measurement
- Documenting traceability

With regard to the function that is played in a measurement process, RMs may be divided into pure substances, those that have a high and strictly defined level of purity, and standard solutions.

The general classification of RMs is presented in Figure 6.1 [8], and a detailed classification of RMs is presented in Table 6.1 [9].

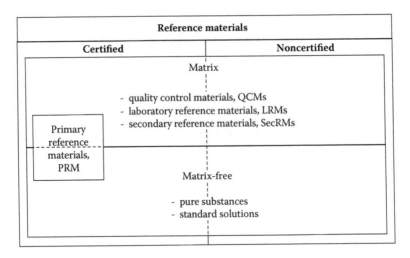

FIGURE 6.1 Classification of reference materials. (From Konieczka, P., *Crit. Rev. Anal. Chem.*, 37, 173–190, 2007.)

Preparation of the RM involves the following:

- Material selection
- Obtaining an appropriate amount of the material
- Selection and purchase of appropriate containers, labels, and so on
- Initial material preparation (grinding, sifting, appropriate fraction grain size)
- Initial examination of the material's homogeneity
- Determination of main components
- Putting the materials into containers
- Final examination of the material's homogeneity
- Disinfection of the material (ensuring its biological stability)
- Determining the humidity
- Organization of an interlaboratory comparison, in order to carry out a certification process
- Statistical analysis of the obtained results (rejection of deviating results; calculating means, standard deviations, and the confidence intervals)
- Determination of values attested to on the basis of hitherto formulated criteria, and printing the attestation certificate

A general procedure for preparing RMs is shown schematically in Figure 6.2 [8].

RM can perform its function only when each of its users receives a material with exactly the same parameters. It may be achieved in two ways: By sending the same material sample, or by sending material samples with the same parameters (homogeneous, stable during storage, stable since the moment of production until their use) [9].

TABLE 6.1
Classification of Reference Materials Suitable for Chemical Investigations

Parameter			Additional Remarks
Property	Chemical composition		Reference materials (RMs), being either pure chemical compounds or representative sample matrices, either natural or with added analytes (e.g., animal fats spiked with pesticides for residue analysis), characterized for one or more chemical or physicochemical property values
	Biological and clinical properties		Materials characterized for one or more biochemical or clinical property values
	Physical properties		Materials characterized for one or more physical property values, that is, melting point, viscosity, density
	Engineering properties		Materials characterized for one or more engineering property values (e.g., hardness, tensile strength or surface characteristics)
	Miscellaneous		These principal categories are subdivided into subcategories as indicated in the following draft list. Other subcategories can be added at any time to address the needs of applicants seeking recognition of competence in producing types of reference materials not currently listed
Chemical nature	Single major constituent	High purity	Pure specific entity (isotope, element, or compound) stochiometrically and isotopically certified in amount-of-substance ratios with total impurities <10 µmol/mol
		Primary chemicals	As above, but with limits of <100 µmol/mol
		Defined purity	As above, but with limits of <50 µmol/mol
	Matrix types	Major constituents	Major constituents (in matrix) >100 mmol/kg or >100 mmol/dm³
		Minor constituents	Minor constituents (in matrix) <100 mmol/kg or <100 mmol/dm³
		Trace constituents	Trace constituents <100 µmol/kg or <100 µmol/dm³
		Ultra trace constituents	Ultra trace constituents <100 nmol/kg or <100 nmol/dm³

(Continued)

TABLE 6.1 (CONTINUED)

Classification of Reference Materials Suitable for Chemical Investigations

Parameter		Additional Remarks
Traceability	0 primary class	Pure specified entity certified to SI at the smallest achievable uncertainty
	I class	Certified by measurement against class 0 RM or SI with defined uncertainty (no measurable dependence on matrix)
	II class	Verified by measurement against class I or 0 RM with defined uncertainty
	III class	Described linkage to class 0, I, II
	IV class	Described linkage other than to SI
	V class	No described linkage
Uncertainty of determination of analyte concentration	With uncertainty value	Primary reference materials (PRMs)
		Certified reference materials (CRMs)
	Without uncertainty value	Laboratory reference materials (LRMs)
		Quality control materials (QCMs)
Field of application		Validation of analytical method
		Establishing measurement traceability
		Calibrating an instrument
		Assessment of a measurement uncertainty
		Assessment of a measurement method
		Recovery studies
		Quality control

Source: Konieczka, P., and Namieśnik, J., Eds., *Kontrola i zapewnienie jakości wyników pomiarów analitycznych*, WNT, Warsaw, 2007 (in Polish).

The selection of the RM depends on the needs at a given time, the type of analytical measurements in which it is going to be used, and its availability.

The differences between noncertified reference materials (laboratory reference material and material for quality control) and certified reference materials (primary reference material and certified reference material) are in accuracy, precision, and the uncertainty of reference value.

That is why CRMs have a higher position in the "metrological hierarchy." The requirements at the production stage, according to ISO recommendations, are more rigorous, which is reflected in their price and thus their availability. Uncertified RMs, including the LRMs (cheaper and more available), are used mainly for the calibration of measuring instruments and checking analytical procedures [9,10].

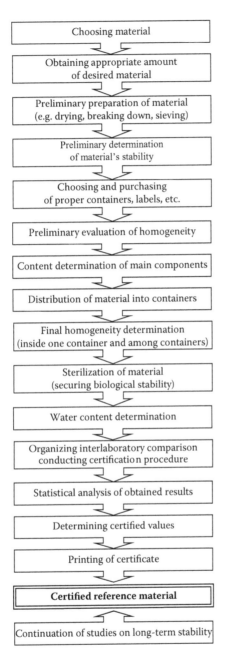

FIGURE 6.2 A general procedure for certified reference materials preparation—example for solid CRMs. (From Konieczka, P., *Crit. Rev. Anal. Chem.*, 37, 173–190, 2007.)

6.3 PARAMETERS THAT CHARACTERIZE RMs

6.3.1 GENERAL INFORMATION

Certification of RMs is something more than just performing a series of accurate and precise measurements traceable to SI standards or to any other metrological system. A certification process involves preparation of a great number of homogeneous, stable, and appropriately packaged samples, which are representative parts of a given production batch.

It is very important to pay special attention not only to the preparation of stable and homogeneous primary materials, but also to sampling [11]. One should take into account microbiological degradation, which can be minimized by decreasing the content of water in the material to the level of 1–3 percent of relative humidity. It is also recommended to pack the RM samples into appropriate containers in the argon atmosphere (bottles with fillers, penicillin vials, or ampoules).

RMs should be prepared in such a way that they are homogeneous, stable, and have constant characteristics over a sufficiently long period.

The parameters that characterize CRMs [12–18] are as follows:

- Representativeness
- Homogeneity
- Stability
- Certified value.

6.3.2 REPRESENTATIVENESS

Representativeness is a property that describes a similarity between individual samples with regard to

- Matrix composition
- Analyte concentration
- Manner of the connection between the analytes and the matrix
- Type and concentration of interfering substances
- Physical state of the material

For practical reasons, achievement of the required similarity is not always possible. A material should be homogeneous and stable, but in the process of homogenization and stabilization a change may occur in the connection between the analyte and the matrix. In such cases, the user should be informed about the actual state of the material, the manner of processing, and how to achieve a representative sample of the material for further analysis.

6.3.3 HOMOGENEITY

Homogeneity study is a comparison of the obtained results for the random samples of the RM. It is carried out at the stage of distributing the RM into the appropriate containers.

There are two types of homogeneity [13]:

- Within-bottle
- Between-bottle

The influence of the *within-bottle* heterogeneity of the material on the result of the certified value may be eliminated by sampling a greater amount of the material. That is why it is necessary to define the minimum amount (mass) of the RM samples for the study.

A user has no influence upon the *between-bottle* heterogeneity of the material. This value should be determined by the producer of the RM and taken into account in the uncertainty budget of the certified value.

Both sources of heterogeneity of RMs are presented in Figure 6.3.

6.3.4 STABILITY

A stability study, next to the homogeneity study, plays a decisive role in the production of RMs. The stability of the RM is determined by using the analysis of

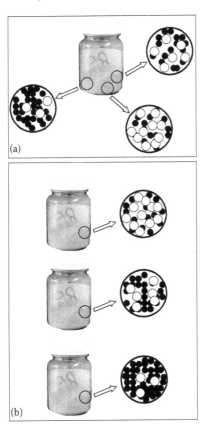

FIGURE 6.3 Sources of reference materials heterogeneity: (a) no within-bottle; (b) between-bottle.

the certified parameters in the samples of materials stored in a so-called reference temperature (with an assumption that in that temperature the composition of the RM does not change) in relation to samples stored in temperatures recommended for a given RM.

During the storage and transportation, the RM is exposed to the influence of various external factors (temperature, light, oxygen, humidity, microbiological activity) that may affect its composition [16]. However, the value of a given parameter of the material should be stable over the whole validity period.

There are two types of RM stability [14–16]:

- Long-term (e.g., shelf-life)
- Short-term (e.g., stability during transportation)

Stability studies require the application of fast measurement methods, low-mass samples, and the high repeatability of the measurements. The studies are carried out for various temperatures and storage durations.

Studying the stability of RMs may be considered in two aspects:

- Classical (long-term)
- Isochronous

In case of the classical stability study, stability is determined by comparing the results obtained for samples stored in the recommended conditions and for the reference samples, usually stored in a lower temperature, for example, −40°C.

Such studies are carried out a short time before the hitherto determined expiry date and may result in extending the validity period.

An isochronous stability study is based on deducing the stability of the RM on the basis of analyses of samples stored over a short period (several weeks) and at various temperatures (usually higher than the recommended storage temperature) [16].

6.3.5 CERTIFIED VALUE

RM certification is carried out according to the strictly determined rules, as described in the appropriate ISO Guides [19–23]. In contrast to pure substances and calibration solutions, matrix RMs cannot be certified using direct gravimetric measurement. In this case, an additional stage is required: a complete change or the removal of the matrix. Thus the following solutions are applied [24]:

- Measurements at a single laboratory, using the absolute methods, that is, methods that give the results directly in units of measurement or methods that allow the result to be expressed in those units through the application of mathematical equations from the appropriate physical and chemical theories

- Measurements at a single laboratory using two or more methods, by two or more analysts
- Interlaboratory studies using one or several various methods, including the absolute methods. It must be remembered that certification studies should be carried out by the laboratories with supreme and proven competence

Certification is based on material sample analyses, using one or more methods at one or several laboratories, in which each of the measurement series is carried out with the highest accuracy and traceability, and must be documented by a complete uncertainty budget.

The aim of material certification is to ascribe certain values of individual properties to a group or individual units. The reliability of the obtained results of analytical measurements is a self-evident condition, essential for certification [25].

The final uncertainty value of the CRM, according to the guidelines presented by Guide to the Expression of Uncertainty in Measurement [26], should include all the uncertainty sources described in the following equation [27]:

$$u_{CRM} = \sqrt{u_{cert}^2 + u_{bott}^2 + u_{ls}^2 + u_{ss}^2} \qquad (6.1)$$

where

u_{cert}: Uncertainty of determining the certified value
u_{bott}: Uncertainty associated with the within-bottle homogeneity
u_{ls}: Uncertainty associated with the long-term stability
u_{ss}: Uncertainty associated with the short-term stability

6.4 PRACTICAL APPLICATION OF CRMs

These are the main issues associated with the application of the RMs [28–31]:

- Determination of validation parameters—first of all, their precision and accuracy
- Examining the skills of an analyst or a laboratory
- Routine control of precision and accuracy of the performed determinations
- Laboratory accreditation
- The quality control of performance of a given laboratory
- Estimating measurement uncertainty
- Monitoring and ensuring traceability
- Calibration of measuring instruments

It is not possible to prepare appropriate RMs for all the analytical tasks, due to the high heterogeneity of matrix compositions and the wide spectrum of analytes present in the examined samples. A good knowledge of analytical procedures and the available materials is, therefore, a key to the right choice.

The selection of the RM should allow for the following criteria:

- Availability (the issue of the matrix composition)
- Concentration range of the reference value
- Uncertainty value of the reference value
- Traceability of the reference value
- Required uncertainty value of the measurement
- Influence of the CRM uncertainty on the combined uncertainty of the measurement
- Quality of the CRM producer (competence, reputation)
- Composition of the sample matrix
- Price

Detailed information concerning the RMs, and help in finding an appropriate RM, can be found in the data bases available at the following websites:

- http://www.comar.bam.de
- http://www.virm.net
- https://nucleus.iaea.org/rpst/ReferenceProducts/About/index.htm
- https://ec.europa.eu/jrc/en/scientific-tool/reference-materials-database-and -online-catalogue
- https://www.nist.gov/srm
- http://www.bipm.org/jctlm

Using RMs requires compliance with the rules of good laboratory practice at laboratories that determine the trace components in the examined samples:

- It is necessary to comply with the recommendations of the RM producer, for example, concerning the minimum mass of the RM sampled, the validity period, and the manner of storage.
- It is necessary to determine the concentration of water (in case of solid materials) for the RM samples taken simultaneously with the RM sample for the study.
- The taken and nonused RM cannot be replaced into containers.

RMs are an essential tool for the determination of accuracy or precision. Because one of the main problems associated with this process is the interpretation and numerical presentation of the determined parameter, this book presents the basic formulas and correlations that help in selecting the manner of documenting the values of the determined parameters.

It seems practical to provide a graphical comparison of the reference (certified) value with the value obtained during the measurement (determined one). Possible situations, depending on the information on the two compared values, together with the associated conclusions are presented in Table 6.2.

TABLE 6.2
A Suitable Way of Graphically Comparing the Reference (Certified) Value with the Determined Value

Conditions	Graphical Presentation	Conclusions
Reference value without providing the uncertainty (not a certified value) and determined value with a provided uncertainty	Reference value · Determined value	Determined value agreed with the reference value
		Conclusion impossible
Reference value with the uncertainty and determined value without a provided uncertainty	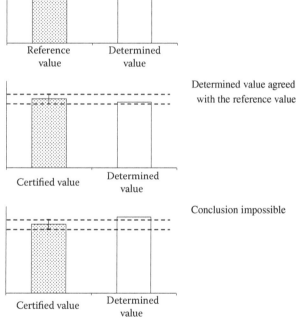 Certified value · Determined value	Determined value agreed with the reference value
		Conclusion impossible

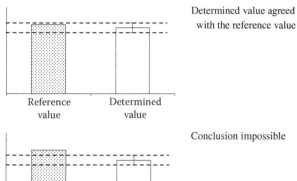
Reference value   Determined value

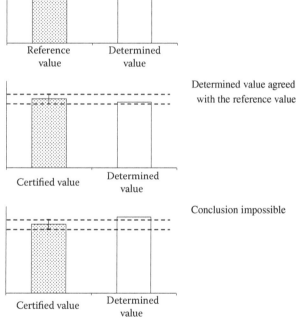
Certified value   Determined value

(*Continued*)

TABLE 6.2 (CONTINUED)

A Suitable Way of Graphically Comparing the Reference (Certified) Value with the Determined Value

Conditions	Graphical Presentation	Conclusions
Reference value with the uncertainty and determined value with a provided uncertainty	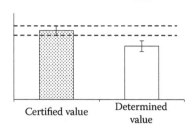	Determined value agreed with the certified value
		Determined value not agreed with the certified value

Certified value   Determined value

Certified value   Determined value

Example 6.1

Problem: Five independent determinations of total mercury were carried out for the samples of the certified reference material NRCC-DORM-2—dogfish muscle.

The certified value given by the producer is 4.64 ± 0.28 µg/g.

Using a graphical method, test the agreement of the obtained value with the certified value.

Data: Result series, µg/g:

1	4.76
2	4.57
3	4.94
4	5.04
5	4.82

Solution:

Mean, µg/g	4.83
SD, µg/g	0.18
U (k = 2), µg/g	0.16

Graph:

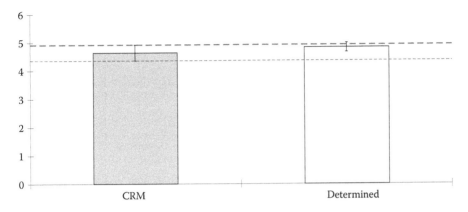

CRM   Determined

Conclusion: An obtained value agreed with certified one.
Excel file: exampl_RM01.xls

Example 6.2

Problem: Six independent determinations of total mercury were carried out for the samples of the reference material GBW 07601—powdered human hair.
The assigned value given by the producer is 0.36 µg/g.
Using a graphical method, test the agreement of the obtained value with the assigned value.
Data: Result series, µg/g:

1	0.38
2	0.34
3	0.35
4	0.39
5	0.37
6	0.40

Solution:

Mean, µg/g	0.372
SD, µg/g	0.023
U (k=2), µg/g	0.019

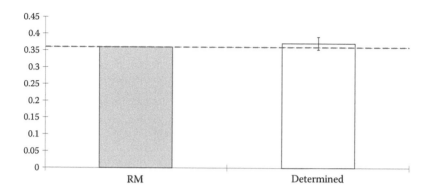

Conclusion: An assigned value is in the range of obtained value ± uncertainty.
Excel file: exampl_RM02.xls

An alternative solution is to determine the conformity of the reference value with the determined value using appropriate tests. The following options are feasible:

1. A comparison of the standard deviation values in the series of measurements for *CRM*, with the value of expanded uncertainty for *CRM*, and the comparison of the determined values with the certified value.
   The following condition must be fulfilled:

$$\frac{SD_{det}}{\sqrt{n}} < U_{CRM} \tag{6.2}$$

where
   SD_{det}: Standard deviation for the measurement series for *CRM*
   n: The number of measurements for *CRM*
   U_{CRM}: The expanded uncertainty for *CRM*

and

$$x_{CRM} - U_{CRM} < x_{det} < x_{CRM} + U_{CRM} \tag{6.3}$$

where
   x_{det}: Determined value
   x_{CRM}: Certified value

Example 6.3

Problem: Five independent determinations of total mercury were carried out for the samples of the certified reference material NRCC-DORM-2—dogfish muscle.
The certified value given by the manufacturer is 4.64 ± 0.28 µg/g.
Using the aforementioned method, test the agreement of the obtained value with the certified value.
Data: Result series, µg/g:

1	4.76
2	4.57
3	4.94
4	5.04
5	4.82

Solution:

x_{det}, µg/g	4.83	
SD_{det}, µg/g	0.18	$\dfrac{SD_{det}}{\sqrt{n}} < U_{CRM}$
n	5	
$\dfrac{SD_{det}}{\sqrt{n}}$, µg/g	0.080	
U_{CRM}, µg/g	0.28	
x_{CRM}, µg/g	4.64	$x_{CRM} - U_{CRM} < x_{det} < x_{CRM} + U_{CRM}$
$x_{CRM} - U_{CRM}$, µg/g	4.36	
$x_{CRM} + U_{CRM}$, µg/g	4.92	

Conclusion: An obtained value agreed with certified one.
Excel file: exampl_RM03.xls

Example 6.4

Problem: Four independent determinations of lead were carried out for the samples of the certified reference material NIST-SRM 1633b—coal fly ash.
The certified value given by the producer is 68.2 ± 1.4 µg/g.
Using the aforementioned method, test the agreement of the obtained value with the certified value.

Data: Result series, µg/g:

1	70.2
2	71.4
3	69.8
4	70.6

Solution:

x_{det}, µg/g	70.5	
SD_{det}, µg/g	0.68	
n	4	
$\dfrac{SD_{det}}{\sqrt{n}}$, µg/g	0.34	$\dfrac{SD_{det}}{\sqrt{n}} < U_{CRM}$
U_{CRM}, µg/g	1.4	
x_{CRM}, µg/g	68.2	
$x_{CRM}-U_{CRM}$, µg/g	66.8	$x_{CRM} - U_{CRM} < x_{det} < x_{CRM} + U_{CRM}$
$x_{CRM} + U_{CRM}$, µg/g	69.6	

Conclusion: An obtained value not agreed with certified one.
Excel file: exampl_RM04.xls

2. Application of Student's t test.
   The value of the parameter t is calculated according to the following formula

$$t = \frac{|x_{det} - x_{CRM}|}{SD_{det}}\sqrt{n} \qquad (6.4)$$

The calculated value should be compared with the critical value from the distribution values for an appropriate significance level (α) and the number of degrees of freedom $f = n - 1$.
   Equation 6.4 does not allow for the uncertainty of the certified value; that is why it is recommended to use its modified version:

$$t = \frac{|x_{det} - x_{CRM}|}{\sqrt{u^2_{(x_{det})} + u^2_{(x_{CRM})}}}\sqrt{n} \qquad (6.5)$$

where
   $u_{(x_{det})}$: Combined uncertainty of the determined value
   $u_{(x_{CRM})}$: Combined uncertainty of the certified value

3. The comparison of the certified value with the determined value, using the uncertainty values for both the values.

The following correlations are examined:

$$\left| x_{det} - x_{CRM} \right| < 2\sqrt{u_{(x_{det})}^2 + u_{(x_{CRM})}^2} \qquad (6.6)$$

$$\left| x_{det} - x_{CRM} \right| \geq 2\sqrt{u_{(x_{det})}^2 + u_{(x_{CRM})}^2} \qquad (6.7)$$

Satisfying the first relation implies conformity of the determined value with the certified value, and satisfying the second relation denotes the lack of conformity between these values.

Example 6.5

Problem: Five independent determinations of total mercury were carried out for the samples of the certified reference material NRCC-DORM-2—dogfish muscle.
The certified value given by the manufacturer is 4.64 ± 0.28 µg/g.
Using the aforementioned method, test the agreement of the obtained value with the certified value.
Data: Result series, µg/g:

1	4.76
2	4.57
3	4.94
4	5.04
5	4.82

Solution:

x_{det}	4.83	
SD_{det}	0.18	
n	5	
$u_{(x_{det})}$	0.080	$\left\| x_{det} - x_{CRM} \right\| < 2\sqrt{u_{(x_{det})}^2 + u_{(x_{CRM})}^2}$
$u_{(x_{CRM})}$	0.14	
$\lvert x_{det} - x_{CRM} \rvert$	0.19	$\left\| x_{det} - x_{CRM} \right\| \geq 2\sqrt{u_{(x_{det})}^2 + u_{(x_{CRM})}^2}$
$2\sqrt{u_{(x_{det})}^2 + u_{(x_{CRM})}^2}$	0.32	

Conclusion: An obtained value agreed with certified one.
Excel file: exampl_RM05.xls

Example 6.6

Problem: Four independent determinations of lead were carried out for the samples of the certified reference material NIST-SRM 1633b—coal fly ash.

The certified value given by the producer is 68.2 ± 1.4 µg/g.

Using the aforementioned method, test the agreement of the obtained value with the certified value.

Data: Result series, µg/g:

1	70.2
2	71.4
3	69.8
4	70.6

Solution:

x_{det}	70.5	
SD_{det}	0.68	
n	4	
$u_{(x_{det})}$	0.34	$\left\|x_{det} - x_{CRM}\right\| < 2\sqrt{u^2_{(x_{det})} + u^2_{(x_{CRM})}}$
$u_{(x_{CRM})}$	0.70	
$\left\|x_{det} - x_{CRM}\right\|$	2.3	$\left\|x_{det} - x_{CRM}\right\| \geq 2\sqrt{u^2_{(x_{det})} + u^2_{(x_{CRM})}}$
$2\sqrt{u^2_{(x_{det})} + u^2_{(x_{CRM})}}$	1.6	

Conclusion: An obtained value not agreed with the certified one.

Excel file: exampl_RM06.xls

4. The application of Z-score.

The value of the Z-score is calculated using the following formula:

$$Z = \frac{x_{det} - x_{CRM}}{s} \tag{6.8}$$

where

s: The value of a deviation unit, which can be calculated as the combined uncertainty of the certified value and the determined value.

The reasoning is carried out using the following relations:

- If |Z| ≤ 2, then the determined value agreed with the reference value.
- If |Z| > 2, then the determined value did not agree with the reference value.

Trueness value, due to application of *CRMs*, can be presented as recovery and should be calculated according to the following equations:

$$\%R = \frac{x_{det}}{x_{CRM}} [\%] \tag{6.9}$$

$$U = k \cdot \frac{\sqrt{\left(u_{(x_{det})}^2 + u_{(x_{CRM})}^2\right)}}{\left(\dfrac{x_{det} + x_{CRM}}{2}\right)} [\%] \tag{6.10}$$

The reasoning should be based on the following:

If the range %R ± U includes the expected 100 percent value, the calculated value of trueness is acceptable.

The value of trueness is usually given as

$$Trueness = \%R \pm U \tag{6.11}$$

and most frequently is expressed in %.

Example 6.7

Problem: Five independent determinations of total mercury were carried out for the samples of the certified reference material NRCC-DORM-2—dogfish muscle.

The certified value given by the manufacturer is 4.64 ± 0.28 µg/g.

Using the obtained result, calculate trueness as a recovery value for $k = 2$.

Data: Result series, µg/g:

1	4.76
2	4.57
3	4.94
4	5.04
5	4.82

Solution:

x_{det}	4.83	
x_{CRM}	4.64	
SD_{det}	0.18	
n	5	
$u_{(x_{det})}$	0.080	$\%R = \dfrac{x_{det}}{x_{CRM}} \cdot 100\%$
$u_{(x_{CRM})}$	0.14	
k	2	
$\%R$	104.0%	$U = k \cdot \dfrac{\sqrt{\left(u_{(x_{det})}^2 + u_{(x_{CRM})}^2\right)}}{\left(\dfrac{x_{det} + x_{CRM}}{2}\right)}$
U	6.8%	

Conclusion: A value of 100 percent is in the range of the calculated trueness value.
Excel file: exampl_RM07.xls

Due to a limited number of certified reference materials, a widely known standard addition method is applied as an alternative manner of determining trueness.

The recovery is calculated based on increasing the signal (recalculated for concentration, content) after standard addition.

It is very important to fulfill requirements for that method, so increasing the signal should be more than 50 percent of the value for sample and less than 150 percent of that value. The volume of the standard added should be negligible compare to the sample volume (no influence on matrix composition).

Example 6.8

Problem: Standard addition method has been used for the determination of trueness. Two series were conducted—for real samples and for samples with standard addition.

Using the obtained result, calculate trueness as a recovery value for $k = 2$. Assume the value $\alpha = 0.05$.
Data: Results series, mg/dm³:

	Data	
	Sample	**Sample with Standard Addition**
1	33.54	57.03
2	33.11	58.11

3	32.87	59.03
4	33.75	57.88
5	34.39	58.23
6	33.33	60.34
7	32.05	57.99

			U	k
Standard Concentration x_{st}	5000	mg/dm³	5	2
Standard Volume V_{st}	0.50	cm³	0.02	2
Sample Volume V_{smpl}	100.0	cm³	0.2	2

Solution:
Checking for outliers, using Dixon's Q test.

	Sample	Sample with Standard Addition
No. of results—n	7	7
Range—R	2.34	3.31
Q_1	0.350	0.257
Q_n	0.274	0.396
Q_{crit}	0.507	0.507

According to the equation from Section 1.8.3:
Because Q_1 and $Q_n < Q_{crit}$, for both series, there are no outliers in the results series. The calculated values of x_m, SD, CV, and $u_{r(det)}$:

	Sample	Sample with Standard Addition	
X_m	33.29	58.37	mg/dm³
SD	0.73	1.0	mg/dm³
CV	2.2	1.8	%
$u_{r(det)}$	0.83	0.68	%

where $u_{r(det)}$ has been calculated as

$$u_{r(det)} = \frac{CV}{\sqrt{n}}$$

The theoretical concentration after standard addition has been calculated according to the following formula

$$X_{teor} = \frac{X_{m(smpl)} \times V_{smpl} + X_{st} \times V_{st}}{V_{smpl} + V_{st}}$$

Theoretical concentration after standard addition x_{theor}	58.00	mg/dm³

The calculations of concentration increasing are

$$\Delta x_{theor} = x_{theor} - x_{smpl}$$

$$\Delta x_{det} = x_{smpl+st} - x_{smpl}$$

Concentration Increasing		
Theoretical Δx_{theor}	Determined Δx_{det}	
24.71	25.08	mg/dm³

Before calculating recovery it is necessary to check if amount of standard added fulfilled a requirement for application of standard addition method. For that both relations have to be fulfilled:

$$0.5 \times x_{det} < \Delta x_{theor} < 1.5 \times x_{det}$$

For the data:

$$16.65 < 24.71 < 49.93$$

Recovery is calculated as

$$\%R = \frac{\Delta x_{det}}{\Delta x_{theor}}$$

And its expanded uncertainty for the value for $k = 2$ is according to the following formula

$$U(k = 2) = 2 \cdot \%R \cdot \sqrt{u^2_{r(det)smpl} + u^2_{r(det)smpl+st} + \left(\frac{U_{x_{st}}}{\frac{k}{x_{st}}}\right)^2 + \left(\frac{U_{V_{st}}}{\frac{k}{V_{st}}}\right)^2 + \left(\frac{U_{V_{smpl}}}{\frac{k}{V_{smpl}}}\right)^2}$$

$\%R$	101.5%
$U(k=2)_{\%R}$	4.6%

A value of 100 percent is in the range of calculated trueness value; there is no need to correct the results on bias.

Conclusion: The investigated method is accurate.

Excel file: exampl_RM08.xls

Example 6.9

Problem: The standard addition method has been used for the determination of trueness. Two series were conducted—for a real sample and for a sample with standard addition.

Using the obtained result, calculate trueness as a recovery value for $k = 2$. Assume the value $\alpha = 0.05$.

Data: Results series, mg/dm³:

	Data	
	Sample	Sample with Standard Addition
1	53.23	110.1
2	54.87	111.6
3	55.98	108.1
4	51.34	121.5
5	50.21	118.1
6	56.11	109.9
7	53.88	115.3

			U	k
Standard concentration x_{st}	5000	mg/dm³	5	2
Standard volume V_{st}	1.30	cm³	0.02	2
Sample volume V_{smpl}	100.0	cm³	0.2	2

Solution:
Checking for outliers, using Dixon's Q test.

	Sample	Sample with Standard Addition
No. of results—n	7	7
Range—R	2.34	3.31
Q_1	0.192	0.134
Q_n	0.022	0.254
Q_{crit}	0.507	0.507

According to the equation from Section 1.8.3:
Because Q_1 and $Q_n < Q_{crit}$, for both series, there are no outliers in the results series. The calculated values of x_m, SD, CV and $u_{r(det)}$:

	Sample	Sample with Standard Addition	
X_m	53.66	113.51	mg/dm³
SD	2.2	4.9	mg/dm³
CV	4.2	4.3	%
$u_{r(det)}$	1.6	1.6	%

where $u_{r(det)}$ has been calculated as

$$u_{r(det)} = \frac{CV}{\sqrt{n}}$$

The theoretical concentration after standard addition has been calculated according to the following formula:

$$X_{teor} = \frac{X_{m(smpl)} \times V_{smpl} + X_{st} \times V_{st}}{V_{smpl} + V_{st}}$$

Theoretical concentration after standard addition x_{theor}	117.14	mg/dm³

The calculations of concentration increasing are

$$\Delta x_{theor} = x_{theor} - x_{smpl}$$

$$\Delta x_{det} = x_{smpl+st} - x_{smpl}$$

Concentration Increasing		
Theoretical Δx_{theor}	Determined Δx_{det}	
63.48	59.85	mg/dm³

Before calculating recovery it is necessary to check if the amount of standard added fulfilled a requirement for application of the standard addition method. For that both relations have to be fulfilled:

$$0.5 \times x_{det} < \Delta x_{theor} < 1.5 \times x_{det}$$

For the data:

$$26.83 < 63.48 < 80.49$$

Recovery is calculated as

$$\%R = \frac{\Delta x_{det}}{\Delta x_{theor}}$$

And its expanded uncertainty for value for $k = 2$ according to the following formula

$$U(k=2) = 2 \cdot \%R \cdot \sqrt{u_{r(det)smpl}^2 + u_{r(det)smpl+st}^2 + \left(\frac{U_{x_{st}}}{x_{st}}\right)^2 + \left(\frac{U_{V_{st}}}{V_{st}}\right)^2 + \left(\frac{U_{V_{smpl}}}{V_{smpl}}\right)^2}$$

$\%R$	94.3%
$U(k=2)_{\%R}$	4.5%

A value of 100 percent is out of the range of calculated trueness value; it is necessary to correct the results on bias.

Conclusion: The investigated method is not accurate.

Excel file: exampl_RM09.xls

6.5   CONCLUSION

Production and certification of RM is very costly, which is why application of CRMs is usually limited to the verification of analytical procedures and only in some exceptional cases for calibration (in comparative methods). Due to financial limitations, it is not recommended to use certified reference materials for a routine intralaboratory statistical control, nor in interlaboratory comparisons. It is recommended, however, in competence tests.

CRMs play a crucial role in the system of estimation, monitoring, and ensuring the quality of analytical measurement results. Their application, as noted above, is necessary in any laboratory. However, it must be said that using CRMs at a laboratory does not automatically ensure the obtainment of reliable results. RMs must be applied in a rational way, and do not nullify the remaining elements of the quality system.

RMs should be stored in conditions that guarantee the stability of their composition over the whole period of use.

REFERENCES

1. International Vocabulary of Metrology—Basic and general concepts and associated terms (VIM), Joint Committee for Guides in Metrology, JCGM 200, 2008.
2. Guidelines for the In-House Production of Reference Materials, version 2, LGC/VAM, 1998.
3. Emons H., Linsinger T.P.J., and Gawlik B.M., Reference materials: Terminology and use. Can't one see the forest for the trees?, *Trends Anal. Chem.*, 23(6), 442–449, 2004.
4. Rasberry S.D., Reference materials in the world of tomorrow, *Fresenius J. Anal. Chem.*, 360, 277–281, 1998.
5. Lipp M., Reference materials—An industry perspective, *Accred. Qual. Assur.*, 9, 539–542, 2004.
6. Pauwels J., and Lamberty A., CRMs for the 21st Century: New Demands and Challenges, *Fresenius J. Anal. Chem.*, 370, 111–114, 2001.
7. Majcen N., A need for clearer terminology and guidance in the role of reference materials in method development and validation, *Accred. Qual. Assur.*, 8, 108–122, 2003.
8. Konieczka P., The role of and place of method validation in the quality assurance and quality control (QA/QC) system, *Crit. Rev. Anal. Chem.*, 37, 173–190, 2007.
9. Konieczka P., and Namieśnik J. (eds.), Kontrola i zapewnienie jakości wyników pomiarów analitycznych, WNT, Warsaw, 2007 (in Polish).
10. Fellin P., and Otson R., A test atmosphere generation system for particle-bound PNA: Development and use for evaluation of air sampling methods, *Chemosphere*, 27, 2307–2315, 1993.
11. Kramer G.N., and Pauwels J., The preparation of biological and environmental reference materials, *Mikrochim. Acta.*, 123, 87–93, 1996.
12. Linsinger T.P.J., Pauwels J., Van der Veen A.M.H., Schimmel H., and Lamberty A., Homogeneity and stability of reference materials, *Accred. Qual. Assur.*, 6, 20–25, 2001.
13. Van der Veen A.M.H., Linsinger T., and Pauwels J., Uncertainty calculations in the certification of reference materials. 2. Homogeneity study, *Accred. Qual. Assur.*, 6, 26–30, 2001.
14. Van der Veen A.M.H., Linsinger T.P.J., Lamberty A., and Pauwels J., Uncertainty calculations in the certification of reference materials. 3. Stability study, *Accred. Qual. Assur.*, 6, 257–263, 2001.
15. Pauwels J., Lamberty A., and Schimmel H., Quantification of the expected shelf-life of certified reference materials, *Fresenius J. Anal. Chem.*, 361, 359–361, 1998.
16. Lamberty A., Schimmel H, and Pauwels J., The study of the stability of reference materials by isochronous measurements, *Fresenius J. Anal. Chem.*, 360, 395–399, 1998.
17. Van der Veen A.M.H., and Pauwels J., Uncertainty calculations in the certification of reference materials. 1. Principles of analysis of variance, *Accred. Qual. Assur.*, 5, 464–469, 2000.

18. Van der Veen A.M.H., Linsinger T.P.J., Schimmel H., Lamberty A., and Pauwels J., Uncertainty calculations in the certification of reference materials. 4. Characterisation and certification, *Accred. Qual. Assur.*, 6, 290–294, 2001.
19. ISO Guide 30, Trends and definitions used in connections with reference materials, ISO, Geneva, 1992.
20. ISO Guide 31, Reference materials—Contents of certificates and labels, Geneva, 2000.
21. ISO Guide 34, Quality system guidelines for the production of reference materials, ISO, Geneva, 1996.
22. ISO Guide 35, Certification of reference materials, General and statistical principles, Geneva, 1989.
23. General requirements for the competence of reference material producers, ISO 17034, Geneva, 2016.
24. Uriano G.A., and Gravatt C.C., The role of reference materials and reference methods in chemical analysis, *Crit. Rev. Anal. Chem.*, 6, 361–411, 1977.
25. Linsinger T.P.J., Pauwels J., Schimmel H., Lamberty A., Veen A.M.H., Schumann G., and Siekmann L., Estimation of the CRMs in accordance with GUM: Application to the certification of four enzyme CRMs, *Fresenius J. Anal. Chem.*, 368, 589–594, 2000.
26. ISO, Guide to the Expression of Uncertainty in Measurement (GUM), Geneva, 1993.
27. Pauwels J., Van der Veen A., Lamberty A. and Schimmel H., Evaluation of uncertainty of reference materials, *Accred. Qual. Assur.*, 5, 95–99, 2000.
28. Caroli S., Forte G., and Iamiceli A.L., ICP-AES and ICP-MS quantification of trace elements in the marine macroalga fucus sample, a new candidate certified reference material, *Microchem. J.*, 62, 244–250, 1999.
29. Sutherland R.A., and Tack F.M.G., Determination of Al, Cu, Fe, Mn, Pb and Zn in certified reference materials using the optimized BCR sequential extraction procedure, *Anal. Chim. Acta*, 454, 249–257, 2002.
30. Caroli S., Senofonte O., Caimi S., Robouch P., Pauwels J., and Kramer G.N., Certified reference materials for research in Antarctica: The case of marine sediment, *Microchem. J.*, 59, 136–143, 1998.
31. Dybczyński R., Danko B., and Polkowska-Motrenko H. Some difficult problems still existing in the preparation and certification of CRMs, *Fresenius J. Anal. Chem.*, 370, 126–130, 2001.

7 Interlaboratory Comparisons

7.1 DEFINITIONS [1,2]

Certification study: A study which assigns a reference value to a given parameter (e.g., analyte concentration) in a tested material or a given sample, usually with a determined uncertainty.

Interlaboratory comparisons: Organization, performance, and evaluation of tests on the same or similar test items by two or more laboratories in accordance with predetermined conditions.

Proficiency testing: Determination of laboratory testing performance by means of interlaboratory comparisons.

Method–performance study: Interlaboratory research in which all participants act according to the same protocol and using the same test procedures to determine the characteristic features in a batch of identical test samples.

7.2 INTRODUCTION

Demand for results as a source of reliable analytical information poses new challenges for analytical laboratories: They need to be especially careful in documenting the results and the applied research methods. Ensuring a suitable quality of analytical results is essential due to the negative implications of presenting unreliable measurement results. The way to realize this goal is to implement a suitable quality assurance system at a laboratory through constant monitoring of the reliability of the analytical results and calibration. One of the most crucial means of that monitoring is participation in various interlaboratory studies [3].

Participation in these programs gives a chance for a laboratory to compare its results with those obtained by other laboratories and to prove its competence, which can be especially significant for laboratories with accreditation or those applying for accreditation. Moreover, participation in analytical interlaboratory comparative studies gives a laboratory a chance to search and detect unexpected errors using comparison with external standards and its own previous results, and in the case of error detection, undertake rectifying action [4].

A generalized scheme for conducting interlaboratory studies is shown in Figure 7.1 [5].

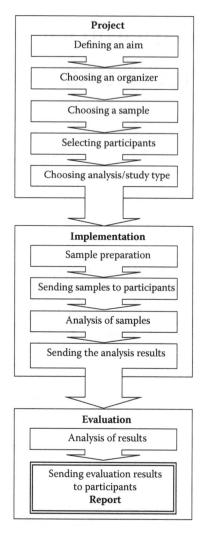

FIGURE 7.1   A generalized outline for conducting interlaboratory studies. (From Konieczka, P., *Crit. Rev. Anal. Chem.*, 37, 173–190, 2007.)

7.3 CLASSIFICATION OF INTERLABORATORY STUDIES

Interlaboratory studies are organized in order to

- Assess the reliability of measurement results
- Gain experience
- Increase the quality of conducted analytical determinations
- Create possibilities for proving the competence of a given laboratory
- Better understand the applied procedures
- Determine validation parameters

Laboratories that wish to confirm their competence should participate in at least one program of interlaboratory research. Accredited laboratories are obliged to provide certificates of participation in such a program, both on a national and international scale.

Interlaboratory comparisons may also be classified according to the aim and range of studies. This may include the following:

- Method performance study
- Competence study
- Certification study
- Proficiency testing

Method performance study is an interlaboratory comparison in which all participants act according to the same protocol and use the same test procedures to determine the characteristic features (specified in the protocol) in a batch of identical test samples. The obtained results are applied in estimating the characteristic parameters of the procedure:

- Intra- and interlaboratory precision
- Systematic error
- Recovery value
- Internal parameters of quality assurance
- Sensitivity
- Limit of detection
- Applicability limit

In this type of research, it is necessary to conform to the following requirements:

- The composition of the applied material or sample is usually similar to that of the materials or samples subjected to routine studies, with regard to the composition of the matrix, analyte concentration, and the presence of interferents (the participants of the research are usually informed about the composition of the matrix for the examined samples).
- The number of participants, test samples, and determinations as well as other details of the study are presented in the research protocol prepared by the organizer of the study.
- By using the same materials or test samples, it is possible to compare a few procedures; all participating laboratories apply the same set of guidelines for each procedure, and the statistical analysis of the obtained sets of results is conducted separately for each of the procedures.

A competence study is a research in which one or more analyses are carried out by a group of laboratories using one or more homogenous and stable test sample and using a selected or routinely used procedure by each of the laboratories participating in the interlaboratory comparison. The obtained sample results are compared with the results obtained by other laboratories or with a known or determined (guaranteed) reference value. This research may be conducted among laboratories that are accredited or applying for accreditation in order to control the quality of determinations and the proficiency of researchers. In this case, the applied analytical procedure may be a top-down decision or the organizer may limit the choice to a prepared list.

A certification study is a study which assigns a reference value to a given parameter (e.g., analyte concentration, physical property) in a tested material or a given sample, usually with a determined uncertainty. This research is usually carried out by laboratories with a confirmed competence (reference laboratories) to test the material, which is a candidate for the reference material, using a procedure that ensures the estimation of the concentration (or any other parameter) with the smallest error and the lowest uncertainty value.

Proficiency testing is the most frequent type of interlaboratory research, which is why it is important to pay it a little more attention. These studies are conducted to test the achievements and competence of both the individual analysts using a given analytical procedure or measurement, and a specific analytical procedure.

Proficiency testing may be conducted on the basis of the same material analysis: sample of the material being provided to all the participants at the same time for a simultaneous study or a round robin test. In the latter case, some problems with the stability and homogeneity of samples may occur due to the spread of the studies over a longer time.

Proficiency testing may be conducted as open (public) studies or as a closed (not public) study. In the case of closed research, the participants do not know that these are proficiency studies and that the obtained samples are to be analyzed in a routine fashion [6].

Proficiency research is a tremendous challenge for laboratories that need to apply for accreditation based on the presentation of confirmation of their own competence. It is a significant element in achieving and maintaining a suitable quality of results. In proficiency testing, the competence of the participating laboratories is verified based on the determination of results of specified components in distributed samples (materials). Each laboratory is assigned an identification number, under which the participant remains anonymous to the rest of the group.

The choice of test material should be influenced by the maximum degree of similarity of the composition of the samples, usually subjected to analysis with regard to the matrix composition and the level of analyte concentration. Such a material must be tested before it is distributed to the participants, with regard to the mean level of analyte concentration and the homogeneity degree. The obtained results are compared with the previously determined guaranteed (assigned) value.

There are six various ways of enabling the determination of the assigned value:

- Measurement by a reference laboratory
- Certified value for CRM used as a test material
- Direct comparison of the PT test material with CRM
- Consensus value from expert laboratories
- Formulation value assignment on the basis of proportions used in a solution or other mixture of ingredients with known analyte contents
- Consensus value from participating laboratories

Sometimes pilot studies are implemented to select the participants with suitable qualification to participate in the actual proficiency studies, the so-called key comparisons. After the initial research, all the participants gather to discuss the obtained results. In the case of results distinctly deviating from the assumed range of acceptable results, the participants try to find the causes of the discrepancies. It gives laboratories a chance to improve their competence, correct the hitherto existing mistakes, and improve their performance in the next proficiency test.

With regard to conditions, there are two main types of proficiency studies:

- Those examining the competence of the group of laboratories using the results from specifically defined types of analyses
- Those examining the competence of laboratories during the performance of various types of analyses

Taking into consideration the sample preparation used by the participating laboratories, each of the aforementioned types may be divided into three further categories:

- Samples circulate successively from one laboratory to another. In this case a sample may be taken back to the coordinating laboratory before a test by a subsequent participant to check if the sample has changed in an undesirable fashion.
- Subsamples randomly selected from a large batch of homogeneous material or test samples are simultaneously distributed to participating laboratories (the most popular type of proficiency testing).
- Product or material samples are divided into several parts and each participant receives one part of each sample (this type is called the split sample study).

There are certain limitations associated with performance and participation in proficiency testing. First of all, proficiency testing is unusually time consuming. It generally takes a long time before the participants get to know the obtained results. Moreover, the interlaboratory comparisons are retrospective studies, which is why proficiency testing may not affect any decision on quality management. In reality, proficiency testing accounts for only a small percentage of analyses conducted by the laboratories and therefore does not reflect the full picture of routinely performed studies.

7.4 CHARACTERISTICS AND ORGANIZATION OF INTERLABORATORY COMPARISONS

As one can see from this discussion, it is necessary to check the work of individual laboratories because it gives them a chance to estimate the reliability of the analytical results of a given research team. Moreover, a thorough analysis of an analytical process, with the cooperation of a control center, produces a precise localization of sources and causes of errors and hence an improvement in the quality of analytical results. The achievement of these aims requires a painstaking and reliable organization of this research.

Reference materials are a necessary tool to conduct interlaboratory comparisons. Their production and certification is usually very expensive, therefore the use of certified reference materials (CRM) should be limited to the verification of analytical procedures and, in the case of comparative methods, should be limited to the calibration of the control and measuring instruments. Due to economic reasons in interlaboratory comparisons one may effectively use laboratory reference materials (LRM).

All the reference materials should fulfill basic requirements with regard to similarity, homogeneity, and stability over a sufficiently long time. Detailed information on the characteristics, production, and implementation of the reference materials is presented in Chapter 6.

7.5   THE PRESENTATION OF INTERLABORATORY COMPARISON RESULTS: STATISTICAL ANALYSIS IN INTERLABORATORY COMPARISONS

The first stage of interlaboratory research result processing is the graphical presentation of the results [7–10]. To this end, a graph may be constructed where the results are marked from the lowest to the highest, assigning each result a code corresponding to the code number of the laboratory. Diagrams of this type are usually presented in final reports by the organizers of interlaboratory comparisons and proficiency tests. The diagrams make it possible for participants to see how their results relate to the results provided by the other participants. They are also a precious source of information for a potential customer or the accreditation office. On the X-axis, laboratory codes are marked, or the applied procedures, and (optionally) the number of performed independent determinations. On the Y-axis, the general mean (or assigned value) is marked, along with the determined uncertainty value, the individual results obtained by the laboratories, and the uncertain values.

Example 7.1

Problem: For a given series of measurement results obtained by various laboratories and a given reference value and its uncertainty, make a diagram showing the distribution of individual determination results.
Data: Results:

	Data	u
lab 1	123	11
lab 2	111.0	9.8
lab 3	128	14
lab 4	138	16
lab 5	121	10
lab 6	123	11
lab 7	188	14
lab 8	114	18

lab 9	188	23
lab 10	122	15
lab 11	121	11
lab 12	142	13
lab 13	125	12
lab 14	132	17
lab 15	129	19
lab 16	121	21
lab 17	198	28
lab 18	131	14
lab 19	158	18
lab 20	193	13
lab 21	122	14
lab 22	111	17

Solution:

x_{ref}	140
u_{ref}	11

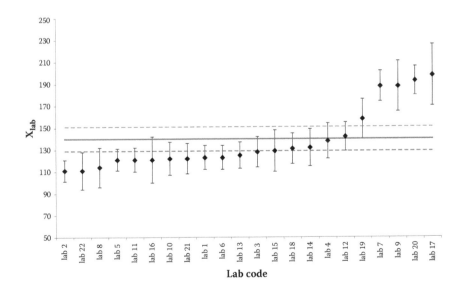

Excel file: exampl_PT01.xls

The manner of conducting a statistical analysis of results obtained in interlaboratory comparisons, and the selection of suitable tests and solutions depend on the type of research. Respective documents define the precise manner of conduct for a specified type of research. The ultimate aim of all types of studies is to determine, based on experimentally obtained numerical data, the accuracy (or precision) of the measurement procedures. On this basis, one may draw conclusions on the applied procedure and on the characteristics of the analyst, compare various procedures, and conduct certification of the material or validation of a specified procedure.

The accuracy of a given measurement procedure may be determined by comparing the assumed reference/assigned value with the mean value of results obtained using the said procedure. Depending on the type of measurements and the requirements for the results, one may use the arithmetical mean or median (parameters presented and defined in Chapter 1).

Precision is associated with the conformity of the series of results. In recording the variability of the results obtained using a given procedure, there are two useful methods of describing precision: Repeatability and reproducibility of results obtained using the specified analytical procedures.

At the initial processing of data provided by the participants of interlaboratory comparisons, the distribution type is examined. The normality of the distribution may be examined using, for example, a Kolmogorov–Smirnov test (Section 1.8.18).

The next step in statistical analysis is to eliminate any deviating results. One checks if the occurrence of doubtful or deviating values may be explained by technical errors. A large number of doubtful or deviating values (outliers) may suggest a significant discrepancy of the variance values or significant differences in the competence between individual laboratories participating in the project, or, lastly, may question the suitability of the selected measurement procedure.

Eliminating the outliers is especially crucial in a situation where the material used in the interlaboratory research is a material for which the reference value is determined based on the results of the very research, for example, when it is a certification study, or when the subject of the comparisons is not the reference material.

To this end, one may use the statistical tests of Cochran (Section 1.8.12) and Grubbs (Section 1.8.13) [11], or the Hampel test (Section 1.8.14), also called the Huber test [9,11]. The choice of a suitable test is conditioned by many factors, first of all, the number of results. There are many reports in which authors critically examined, analyzed, and compared various test used for outlier rejection.

Example 7.2

Problem: Find outliers in a given series of measurement results obtained by various laboratories using Hampel's test.
Data: Results:

	Data
lab 1	123
lab 2	111
lab 3	128
lab 4	138

lab 5	121
lab 6	123
lab 7	188
lab 8	114
lab 9	188
lab 10	122
lab 11	121
lab 12	142
lab 13	125
lab 14	132
lab 15	129
lab 16	121
lab 17	198
lab 18	131
lab 19	158
lab 20	193
lab 21	122
lab 22	111

Solution:

	$\lvert r_i \rvert$	Data	Outlier or not
lab 1	3.5	123	OK
lab 2	15.5	111	OK
lab 3	1.5	128	OK
lab 4	11.5	138	OK
lab 5	5.5	121	OK
lab 6	3.5	123	OK
lab 7	61.5	188	outlier
lab 8	12.5	114	OK
lab 9	61.5	188	outlier
lab 10	4.5	122	OK
lab 11	5.5	121	OK
lab 12	15.5	142	OK
lab 13	1.5	125	OK
lab 14	5.5	132	OK
lab 15	2.5	129	OK
lab 16	5.5	121	OK
lab 17	71.5	198	outlier
lab 18	4.5	131	OK
lab 19	31.5	158	outlier
lab 20	66.5	193	outlier
lab 21	4.5	122	OK
lab 22	15.5	111	OK

SD	8.5	after outlier rejected
X_m	124.4	

Excel file: exampl_PT02.xls

Example 7.3

Problem: Find outliers in the given sets of measurement results obtained in interlaboratory comparisons. Use the Cochran test to examine the intralaboratory variability.
Data: Results:

lab 1	12.1	12.6	13.4
lab 2	11.8	12.0	11.4
lab 3	12.8	14.1	13.5
lab 4	11.8	12.1	13.1
lab 5	11.4	10.9	11.0
lab 6	12.6	11.5	13.1
lab 7	13.6	14.1	12.6
lab 8	14.1	12.8	13.7

Solution:

	Mean	**SD**	**V**
lab 1	12.70	0.66	0.430
lab 2	11.73	0.31	0.093
lab 3	13.47	0.65	0.423
lab 4	12.33	0.68	0.463
lab 5	11.10	0.26	0.070
lab 6	12.40	0.82	0.670
lab 7	13.43	0.76	0.583
lab 8	13.53	0.67	0.443

n	3
p	8
C	0.211
$C_{0.05}$	0.516
$C_{0.01}$	0.615

Conclusion: The result obtained by "lab 6" is correct.
Excel file: exampl_PT03.xls

Example 7.4

Problem: Find outliers in the given sets of results obtained in interlaboratory comparisons from Example 7.3. Apply Grubbs' test for one outlier to examine the interlaboratory variability.
Data: Results:

lab 1	12.1	12.6	13.4
lab 2	11.8	12.0	11.4
lab 3	12.8	14.1	13.5
lab 4	11.8	12.1	13.1
lab 5	11.4	10.9	11.0
lab 6	12.6	11.5	13.1
lab 7	13.6	14.1	12.6
lab 8	14.1	12.8	13.7

Solution:

	Mean
lab 1	12.70
lab 2	11.73
lab 3	13.47
lab 4	12.33
lab 5	11.10
lab 6	12.40
lab 7	13.43
lab 8	13.53

n	3
p	8
x_m	12.588
SD	0.881
G_p	1.688
min/max	MIN
$G_{0.01}$	2.274
$G_{0.05}$	2.126

Conclusion: Result obtained by "lab 5" is correct.
Excel file: exampl_PT04.xls

To simultaneously determine the standard deviation as the measures of repeatability and reproducibility, one may perform a one-factor (one-dimensional) analysis of variance (ANOVA). This analysis serves to verify the hypothesis that the means in the groups are identical against the alternative hypothesis (at least two means are different).

The obtained numerical data are divided into m groups, according to their origin (m is the number of laboratories). When significant differences are found between the values of random errors (statistically significant differences in the variance values), the data are joined into groups for which the variance values are not statistically significantly different, and then the variance analysis is conducted for each group.

An essential condition for conducting a correct interpretation of results for this analysis is the normal distribution of the population from which the samples were taken, with the identical value of the variance V. The essence of the variance analysis is the division of the total variability, that is, the total sum of the squared deviations from all the measurement from the mean, by the sum of squares describing the variability within groups and the sum of squares describing the variability among groups. Then one should determine the total intra- and intergroup degrees of freedom and calculate the standard deviation within individual groups and among the groups, the standard deviation being the measure of the respective variances.

The reliability of conclusions depends, to a great extent, on the number of laboratories participating in the research. Below four degrees of freedom, the value of the parameter $t(\alpha, f)$ increases considerably and the precision of the evaluation decreases. It shows that the interlaboratory studies should involve at least five laboratories. The lower influence on the size of the certainty range is exerted by the number of parallel analyses conducted at a given laboratory. The number of parallel determinations that is greater than five occurs only in special cases, or when for some reason one expects deviation of the obtained measurement results from the normal distribution.

Situations in which a single factor completely explains a given phenomenon are rare. A total error, characterizing the results obtained by using an analytical procedure, consists of a few errors which are summed up according to the law of error propagation.

The parameter used most often to evaluate the obtained results in interlaboratory comparisons is the Z-score parameter. The manner of calculating this parameter has been described in detail in Chapter 1 (Section 1.8.15). The numerical value of the Z-score parameter depends on the number and the type of data available to an analyst:

- When only the mean values obtained from the participating laboratories are known, the assigned (reference) values and the standard deviation sample are calculated according to all the results as the mean value and standard deviations, of course, after rejecting the outliers.

Example 7.5

Problem: In the series of measurement results given in Example 7.1, find which results are satisfactory, which are questionable, and which are unsatisfactory. Use the Z-score. Draw a graph with Z-score values for each of the laboratories.

Data: Results:

	Data
lab 1	123
lab 2	111
lab 3	128
lab 4	138
lab 5	121
lab 6	123
lab 7	188
lab 8	114
lab 9	188
lab 10	122
lab 11	121
lab 12	142
lab 13	125
lab 14	132
lab 15	129
lab 16	121
lab 17	198
lab 18	131
lab 19	158
lab 20	193
lab 21	122
lab 22	111

Solution:

	z	Conclusion
lab 1	−0.16	satisfactory
lab 2	−1.58	satisfactory
lab 3	0.43	satisfactory
lab 4	1.61	satisfactory
lab 5	−0.40	satisfactory
lab 6	−0.16	satisfactory
lab 7	7.51	unsatisfactory
lab 8	−1.22	satisfactory
lab 9	7.51	unsatisfactory
lab 10	−0.28	satisfactory
lab 11	−0.40	satisfactory
lab 12	2.08	questionable
lab 13	0.08	satisfactory
lab 14	0.90	satisfactory
lab 15	0.55	satisfactory

lab 16	−0.40	satisfactory
lab 17	8.69	unsatisfactory
lab 18	0.78	satisfactory
lab 19	3.97	unsatisfactory
lab 20	8.10	unsatisfactory
lab 21	−0.28	satisfactory
lab 22	−1.58	satisfactory

x_m	124.4
SD	8.5

Graph:

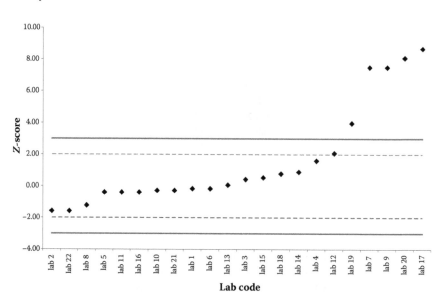

Excel file: exampl_PT05.xls

- Known mean values obtained by the participating laboratories and known assigned/reference value—the value of standard deviation is calculated according to the total set of measurement results—obviously after rejecting the outliers.

Example 7.6

Problem: In the series of measurement results given in Example 7.1, find for a given reference value which results are satisfactory, which are questionable, and which are unsatisfactory. Use the Z-score. Draw a graph with Z-score values for each of the laboratories.

Data: Results:

	Data
lab 1	123
lab 2	111
lab 3	128
lab 4	138
lab 5	121
lab 6	123
lab 7	188
lab 8	114
lab 9	188
lab 10	122
lab 11	121
lab 12	142
lab 13	125
lab 14	132
lab 15	129
lab 16	121
lab 17	198
lab 18	131
lab 19	158
lab 20	193
lab 21	122
lab 22	111

x_{ref}	140

Solution:

SD	8.5

	Z	Conclusion
lab 1	−2.01	questionable
lab 2	−3.42	unsatisfactory
lab 3	−1.42	satisfactory
lab 4	−0.24	satisfactory
lab 5	−2.24	questionable
lab 6	−2.01	questionable
lab 7	5.67	unsatisfactory
lab 8	−3.07	unsatisfactory
lab 9	5.67	unsatisfactory
lab 10	−2.13	questionable
lab 11	−2.24	questionable
lab 12	0.24	satisfactory
lab 13	−1.77	satisfactory
lab 14	−0.94	satisfactory
lab 15	−1.30	satisfactory
lab 16	−2.24	questionable
lab 17	6.85	unsatisfactory
lab 18	−1.06	satisfactory
lab 19	2.13	questionable
lab 20	6.26	unsatisfactory
lab 21	−2.13	questionable
lab 22	−3.42	unsatisfactory

Graph:

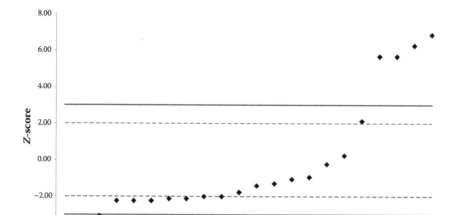

Excel file: exampl_PT06.xls

- Known mean values obtained by the participating laboratories, known assigned/reference value and its combined uncertainty for a given material.

Example 7.7

Problem: In the series of measurement results given in Example 7.1, find which of the results are satisfactory, questionable, or unsatisfactory for the given reference value and the combined uncertainty reference value. Use the Z-score. Draw a graph with the Z-score values for each of the laboratories.

Data: Results:

	Data
lab 1	123
lab 2	111
lab 3	128
lab 4	138
lab 5	121
lab 6	123
lab 7	188
lab 8	114
lab 9	188
lab 10	122
lab 11	121
lab 12	142
lab 13	125
lab 14	132
lab 15	129
lab 16	121
lab 17	198
lab 18	131
lab 19	158
lab 20	193
lab 21	122
lab 22	111

x_{ref}	140
u_{ref}	11

Solution:

	Z	Conclusion
lab 1	−1.55	satisfactory
lab 2	−2.64	questionable
lab 3	−1.09	satisfactory
lab 4	−0.18	satisfactory
lab 5	−1.73	satisfactory
lab 6	−1.55	satisfactory
lab 7	4.36	unsatisfactory
lab 8	−2.36	questionable
lab 9	4.36	unsatisfactory
lab 10	−1.64	satisfactory
lab 11	−1.73	satisfactory
lab 12	0.18	satisfactory
lab 13	−1.36	satisfactory
lab 14	−0.73	satisfactory
lab 15	−1.00	satisfactory
lab 16	−1.73	satisfactory
lab 17	5.27	unsatisfactory
lab 18	−0.82	satisfactory
lab 19	1.64	satisfactory
lab 20	4.82	unsatisfactory
lab 21	−1.64	satisfactory
lab 22	−2.64	questionable

Graph:

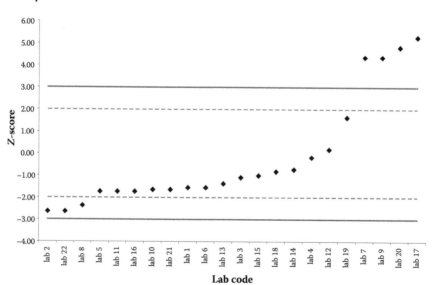

Excel file: exampl_PT07.xls

- Known mean values obtained in the participating laboratories and known value of the reference combined uncertainty for a given material.

Example 7.8

Problem: In a series of measurement results given in the Example 7.1, use the Z-score again, taking into consideration the combined uncertainty reference value. Draw a graph with the Z-score values for each of the laboratories.
Data: Results:

	Data	u
lab 1	123	11
lab 2	111.0	9.8
lab 3	128	14
lab 4	138	16
lab 5	121	10
lab 6	123	11
lab 7	188	14
lab 8	114	18
lab 9	188	23
lab 10	122	15
lab 11	121	11
lab 12	142	13
lab 13	125	12
lab 14	132	17
lab 15	129	19
lab 16	121	21
lab 17	198	28
lab 18	131	14
lab 19	158	18
lab 20	193	13
lab 21	122	14
lab 22	111	17

x_{ref}	140
u_{ref}	11

Solution:

	z	Conclusion
lab 1	−1.09	satisfactory
lab 2	−1.97	satisfactory
lab 3	−0.67	satisfactory
lab 4	−0.10	satisfactory
lab 5	−1.28	satisfactory
lab 6	−1.09	satisfactory
lab 7	2.70	questionable
lab 8	−1.23	satisfactory
lab 9	1.88	satisfactory
lab 10	−0.97	satisfactory
lab 11	−1.22	satisfactory
lab 12	0.12	satisfactory
lab 13	−0.92	satisfactory
lab 14	−0.40	satisfactory
lab 15	−0.50	satisfactory
lab 16	−0.80	satisfactory
lab 17	1.93	satisfactory
lab 18	−0.51	satisfactory
lab 19	0.85	satisfactory
lab 20	3.11	unsatisfactory
lab 21	−1.01	satisfactory
lab 22	−1.43	satisfactory

Graph:

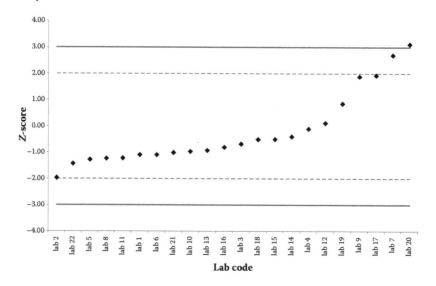

Excel file: exampl_PT08.xls

Another parameter of the individual examination of the measurement results is the relative error. It is calculated in instances when participants of a given study use various methods to evaluate the obtained results, and therefore there is no ground to assume a common value of the sample. It is calculated using the following formula:

$$\varepsilon = \frac{x_{lab} - x_{ref}}{x_{ref}}[\%] \tag{7.1}$$

where
ε: Relative error, %,
x_{lab}: The value of the result obtained by a given laboratory,
x_{ref}: Reference value

Evaluation of the obtained results is obvious in this case and depends on the range of analyte concentrations in a given sample. It is assumed that if

- $|\varepsilon| \leq x$, the evaluation is satisfactory
- $|\varepsilon| > x$, the evaluation is not satisfactory

where x equals relative systematic error (relative deviation), assumed as a limit (permissible).

Example 7.9

Problem: For the data from Example 7.1, calculate the values of the relative errors and make an evaluation for the permissible error value ± 20 percent.
Data: Results:

	Data
lab 1	123
lab 2	111.0
lab 3	128
lab 4	138
lab 5	121
lab 6	123
lab 7	188
lab 8	114
lab 9	188
lab 10	122
lab 11	121
lab 12	142
lab 13	125
lab 14	132
lab 15	129

lab 16	121
lab 17	198
lab 18	131
lab 19	158
lab 20	193
lab 21	122
lab 22	111

x_{ref}	140
$x, \%$	20.0

Solution:

	ε	Conclusion
lab 1	−12.1%	satisfactory
lab 2	−20.7%	unsatisfactory
lab 3	−8.6%	satisfactory
lab 4	−1.4%	satisfactory
lab 5	−13.6%	satisfactory
lab 6	−12.1%	satisfactory
lab 7	34.3%	unsatisfactory
lab 8	−18.6%	satisfactory
lab 9	34.3%	unsatisfactory
lab 10	−12.9%	satisfactory
lab 11	−13.6%	satisfactory
lab 12	1.4%	satisfactory
lab 13	−10.7%	satisfactory
lab 14	−5.7%	satisfactory
lab 15	−7.9%	satisfactory
lab 16	−13.6%	satisfactory
lab 17	41.4%	unsatisfactory
lab 18	−6.4%	satisfactory
lab 19	12.9%	satisfactory
lab 20	37.9%	unsatisfactory
lab 21	−12.9%	satisfactory
lab 22	−20.7%	unsatisfactory

Excel file: exampl_PT09.xls

The next parameter of the individual evaluation (for each of the laboratories) of the obtained results is E_n. The method of its determination is described in detail in Chapter 1 (Section 1.8.16).

E_n is a parameter that is decidedly less restrictive than, for example, the standardized Z coefficient, because of the inclusion of the uncertainty value. Results that are deemed satisfactory may include values significantly deviating from the mean, but within the accepted interval, solely attributable to the high value of the extended uncertainty. An opposite situation is possible—a result closer to the mean (compared with another result from a given series) but with the smaller value of extended uncertainty, may be considered an outlier.

Example 7.10

Problem: For the data from Example 7.1, apply the E_n score.
Data: Results:

	Data	u
lab 1	123	11
lab 2	111.0	9.8
lab 3	128	14
lab 4	138	16
lab 5	121	10
lab 6	123	11
lab 7	188	14
lab 8	114	18
lab 9	188	23
lab 10	122	15
lab 11	121	11
lab 12	142	13
lab 13	125	12
lab 14	132	17
lab 15	129	19
lab 16	121	21
lab 17	198	28
lab 18	131	14
lab 19	158	18
lab 20	193	13
lab 21	122	14
lab 22	111	17

x_{ref}	140
u_{ref}	11

Solution:

	E_n	Conclusion
lab 1	−1.09	unsatisfactory
lab 2	−1.97	unsatisfactory
lab 3	−0.67	satisfactory
lab 4	−0.10	satisfactory
lab 5	−1.28	unsatisfactory
lab 6	−1.09	unsatisfactory
lab 7	2.70	unsatisfactory
lab 8	−1.23	unsatisfactory
lab 9	1.88	unsatisfactory
lab 10	−0.97	satisfactory
lab 11	−1.22	unsatisfactory
lab 12	0.12	satisfactory
lab 13	−0.92	satisfactory
lab 14	−0.40	satisfactory
lab 15	−0.50	satisfactory
lab 16	−0.80	satisfactory
lab 17	1.93	unsatisfactory
lab 18	−0.51	satisfactory
lab 19	0.85	satisfactory
lab 20	3.11	unsatisfactory
lab 21	−1.01	unsatisfactory
lab 22	−1.43	unsatisfactory

Excel file: exampl_PT10.xls

7.5.1 Comparisons of Results Obtained Using Various Procedures

In this type of comparison, box plots may be used. In the graphical presentation of results, one may examine if the results obtained using various analytical procedures differ among themselves in a statistically significant way. In drawing such a plot, one should divide all the measurement results obtained for a given sample into subsets, each containing results obtained using a specific analytical procedure. Then, for each subset, separate plots are drawn, after which they are all put into one diagram.

Based on data for which the diagrams (plots) are drawn, one calculates the essential values based on the following reasoning:

- Ordering the result in a nondecreasing sequence
- Determination of median and quartiles: First (q_1) and third (q_3)
- Determination of the interquartile value (IQR), the difference between q_3 i q_1
- Determination of maximum values, whiskers, as 1.5 times the IQR

Based on these calculated values, a diagram (plot) is drawn (separately for a given set of results) in the following manner:

1. On the OY-axis, for a given series marked by one point on the OX-axis, the values of median and quartiles (q_1 and q_3) are marked—it is a so-called box area representing the middle 50 percent of the data.
2. On the same plot, whiskers are marked as
   (a) $whisker_{min}$, the minimum value in the set of results, not smaller than the limit equal $q_1-1.5 \cdot IQR$; if the so calculated value is equal to q_1, then the $whisker_{min}$ is not marked on the diagram.
   (b) $whisker_{max}$, the maximum value in the set of results, not higher than the limit equal $q_3+1.5 \cdot IQR$; if the so calculated value is equal to q_3, then the $whisker_{max}$ is not marked on the diagram.
3. Results out of this range (lower than $whisker_{min}$ or higher than $whisker_{max}$) are marked as outliers.

Due to that type of construction of the graph, it is possible to conclude which of the analytical procedures were used more often, and which procedure yields more accurate data.

Example 7.11

Problem: For the data from Example 7.1, construct a box plot graph.
Data: Results:

	Data	u
lab 1	123	11
lab 2	111.0	9.8
lab 3	128	14
lab 4	138	16
lab 5	121	10
lab 6	123	11
lab 7	188	14
lab 8	114	18
lab 9	188	23
lab 10	122	15
lab 11	121	11
lab 12	142	13
lab 13	125	12
lab 14	132	17
lab 15	129	19
lab 16	121	21
lab 17	198	28
lab 18	131	14
lab 19	158	18
lab 20	193	13
lab 21	122	14
lab 22	111	17

Solution:

Median—Me	126.5
q_1	121.3
q_3	141.0
IQR	19.8
$1.5 \times IQR$	29.6
$q_1 - 1.5 \times IQR$	91.6
$q_3 + 1.5 \times IQR$	170.6
min	111
max	198
$whisker_{min}$	111
$whisker_{max}$	158

	x_{lab}	Outlier or not
lab 1	123	OK
lab 2	111	OK
lab 3	128	OK
lab 4	138	OK
lab 5	121	OK
lab 6	123	OK
lab 7	188	outlier
lab 8	114	OK
lab 9	188	outlier
lab 10	122	OK
lab 11	121	OK
lab 12	142	OK
lab 13	125	OK
lab 14	132	OK
lab 15	129	OK
lab 16	121	OK
lab 17	198	outlier
lab 18	131	OK
lab 19	158	OK
lab 20	193	outlier
lab 21	122	OK
lab 22	111	OK

Graph:

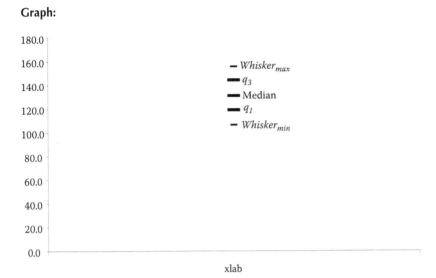

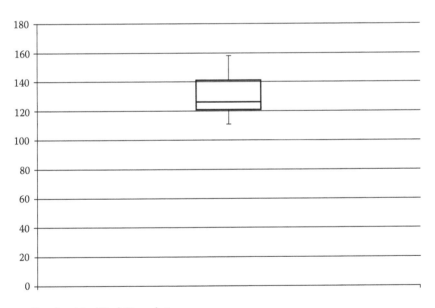

Graph—Modified (Box plot)
Excel file: exampl_PT11.xls

7.5.2   Comparison of the Measurement Results Obtained in a Two-Level Study (for Two Samples with Various Analyte Concentrations)

A two-level study is a study where each of the participating laboratories has performed the series of determinations:

- Either two series per one sample
- Or determinations for two different samples

In this case, to determine the presence of systematic errors, a graphical method—also called the Youden diagram [8] may be used. It is an easy and also very effective method of comparing both intra- and interlaboratory variability. Application of this graph shows which of the participating laboratories achieved comparable results and which laboratory obtained deviating results.

The graph is constructed as follows:

- Measurement results for both the obtained series are marked on the X- and Y-axes.
- Solid lines are drawn (both vertical and horizontal) which reflect the values of the main distribution estimators (arithmetic mean or median).
- Dotted lines are drawn (also vertical and horizontal) where the distances from the solid lines represent values of the standard deviation from the values of the main distribution estimators (arithmetic mean or median.

The distribution of points on such a constructed diagram is a source of information about what type of error has a dominant impact on the obtained measurement results. When the main cause of the deviations from the mean or median are random errors, the results are distributed in a random manner around the mean (median). If a systematic error is the main cause of differences between the values of the measurement results obtained by the compared laboratories and the mean (median), then the majority of points are in the upper-right or bottom-left quarter of the graph. It may indicate a positive or negative bias in the analytical procedure applied in a given laboratory.

Example 7.12

Problem: For the two given series of measurement results for two examined samples obtained in the examining laboratories, produce a Youden graph.
Data: Results:

	Data	
	Series 1	Series 2
lab 1	11.2	12.3
lab 2	10.8	11.8

lab 3	11	12.8
lab 4	10.7	11.7
lab 5	10.5	11.4
lab 6	10.3	11
lab 7	11.2	12.7
lab 8	11.8	13.8
lab 9	12.1	14.2
lab 10	12.9	15.9
lab 11	10.7	11.7
lab 12	11.6	10.9
lab 13	11.4	11.5

Solution:

	Median
Series 1	11.2
Series 2	11.8

Graph:

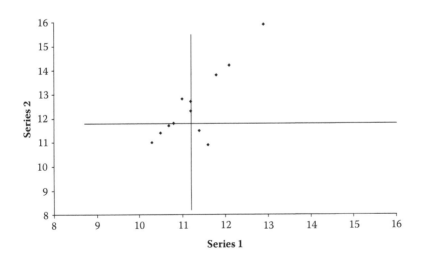

Graph: Modified (with 95 percent limit circle):

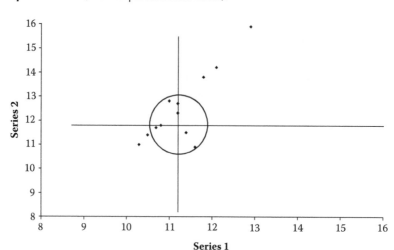

Excel file: exampl_PT12.xls

Another quite common method of graphical presentation of the measurement results obtained by comparing laboratories is the application of Mandel h and k tests. The application of these tests enables the presentation of the variability of results obtained by using a given analytical procedure and enables an evaluation of a given laboratory. The manner of conducting Mandel h and k tests is described in Chapter 1 (Section 1.8.17). All laboratories may obtain on different levels of a study (for different analytes or for different concentrations of a single analyte) both positive and negative values of parameter h.

The number of laboratories characterized with positive values of the parameter h should approximate the number of laboratories characterized with negative values. When a laboratory tends to obtain only negative values for h, one may suppose that there is a source of bias for the results obtained by that laboratory.

Similarly, one should pay attention to a situation where all values of parameter h for a given laboratory are characterized with a positive or negative value, and at the same time different from the sign (plus or minus) of the parameter h obtained in other laboratories.

Moreover, when a laboratory yields h values in the extreme range, for example, it achieved an unusually high number of large values of the h parameter, and the situation should be adequately explained.

When the graph of the statistical parameter k indicates that a given laboratory deviates from the others due to numerous high values, it shows a smaller repeatability of results obtained by the laboratory compared with the rest of the participants. When the graphs of the h and k connected in groups corresponding to the individual laboratories show that the values of these parameters are close to the lines of critical values, one should pay attention to the problem of systematic errors and the small repeatability of results (great variance value).

Example 7.13

Problem: For a given set of results obtained in the interlaboratory comparison, calculate the values of Mandel's h test parameter. Draw a graph showing the respective values of the calculated h parameters characterizing the sets of results obtained in individual laboratories.

Data: Results:

		Level 1	Level 2	Level 3	Level 4	Level 5
	1	4.44	7.21	2.34	14.4	11.3
lab 1	2	4.32	7.54	2.01	15.2	11
	3	4.22	7.77	2.15	13.8	12.5
	1	4.98	7.34	2.03	12.7	10.8
lab 2	2	4.56	7.77	2.12	13.9	11.5
	3	4.73	7.54	2.44	14.2	11.8
	1	5.11	7.67	1.89	14.8	9.96
lab 3	2	5.03	7.83	1.98	16.4	10.4
	3	5.08	7.54	1.78	15.7	10.3
	1	2.22	5.23	1.12	11	6.21
lab 4	2	2.11	5.22	1.45	10.6	6.34
	3	2.34	5.01	1.48	10	6.11
	1	4.56	8.67	2.65	14.5	11.8
lab 5	2	4.76	9.02	2.73	14.2	12.2
	3	4.23	8.92	2.55	14	12
	1	4.11	8.45	2.22	13.3	11
lab 6	2	4.23	8.23	2.86	13.8	11.4
	3	4.02	8.11	2.56	14.1	11.7
	1	4.44	8.11	2.11	13.2	11
lab 7	2	4.55	8.02	2.08	13.1	12
	3	4.21	7.88	2.22	13.6	12.3
	1	3.32	8.98	1.56	11.8	8.76
lab 8	2	3.35	9.11	1.45	11.3	8.67
	3	3.45	9	1.57	11.2	8.82

Solution:

	Level 1	Level 2	Level 3	Level 4	Level 5
			h		
lab 1	0.2516	−0.2088	0.2441	0.6667	0.5948
lab 2	0.7263	−0.1727	0.3104	0.1414	0.4780
lab 3	1.0759	−0.0643	−0.3822	1.3737	−0.0957
lab 4	−2.0704	−2.1712	−1.5611	−1.7172	−2.0970
lab 5	0.4614	0.9280	1.2977	0.5252	0.7949
lab 6	0.0235	0.4221	1.0840	0.2222	0.4780
lab 7	0.3326	0.2053	0.1778	−0.0404	0.6782
lab 8	−0.8008	1.0615	−1.1706	−1.1717	−0.8312

Graph:

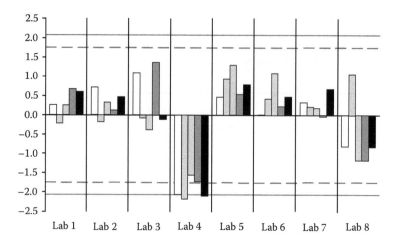

Conclusion: Results obtained by "lab 4" for all the analytes are much lower when compared to those obtained by the rest—three of five analytes have exceeded the critical value for the 1 percent level of significance, which indicates the occurrence of a systematic error source for the results obtained by this laboratory. Results obtained by the other laboratories are within the permissible range of changes for all the determined analytes.
Excel file: exampl_PT13.xls

Example 7.14

Problem: For a given set of results obtained in an interlaboratory comparison, calculate the values of Mandel's k parameter. Draw a graph showing the respective values of the calculated k parameters characterizing the sets of results obtained in individual laboratories.
Data: Results:

		Level 1	Level 2	Level 3	Level 4	Level 5
	1	4.44	7.21	2.34	14.4	11.3
lab 1	2	4.32	7.54	2.01	15.2	11
	3	4.22	7.77	2.15	13.8	12.5
	1	4.98	7.34	2.03	12.7	10.8
lab 2	2	4.56	7.77	2.12	13.9	11.5
	3	4.73	7.54	2.44	14.2	11.8
	1	5.11	7.67	1.89	14.8	9.96
lab 3	2	5.03	7.83	1.98	16.4	10.4
	3	5.08	7.54	1.78	15.7	10.3
	1	2.22	5.23	1.12	11	6.21
lab 4	2	2.11	5.22	1.45	10.6	6.34
	3	2.34	5.01	1.48	10	6.11

	1	4.56	8.67	2.65	14.5	11.8
lab 5	2	4.76	9.02	2.73	14.2	12.2
	3	4.23	8.92	2.55	14	12
	1	4.11	8.45	2.22	13.3	11
lab 6	2	4.23	8.23	2.86	13.8	11.4
	3	4.02	8.11	2.56	14.1	11.7
	1	4.44	8.11	2.11	13.2	11
lab 7	2	4.55	8.02	2.08	13.1	12
	3	4.21	7.88	2.22	13.6	12.3
	1	3.32	8.98	1.56	11.8	8.76
lab 8	2	3.35	9.11	1.45	11.3	8.67
	3	3.45	9	1.57	11.2	8.82

Solution:

	Level 1	Level 2	Level 3	Level 4	Level 5
			k		
lab 1	0.7165	1.6163	0.9477	1.2770	1.7792
lab 2	1.3741	1.2355	1.2330	1.4431	1.1503
lab 3	0.2629	0.8341	0.5732	1.4583	0.5170
lab 4	0.7482	0.7133	1.1430	0.9151	0.2585
lab 5	1.7408	1.0352	0.5160	0.4576	0.4483
lab 6	0.6853	0.9901	1.8323	0.7348	0.7872
lab 7	1.1285	0.6655	0.4218	0.4810	1.5258
lab 8	0.4427	0.4019	0.3810	0.5845	0.1692

Graph:

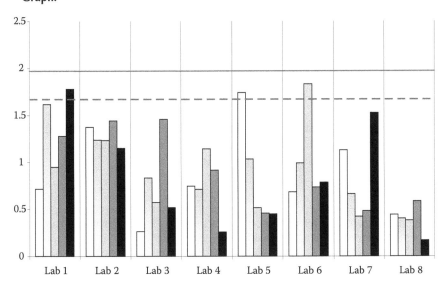

Conclusion: The greatest repeatability for results obtained is achieved by "lab 8."
In the case of individual results ("lab 1," "lab 5," and "lab 6"), the obtained values of repeatability exceed the critical value for the 5 percent level of significance.
Excel file: exampl_PT14.xls

7.6 CONCLUSION

The ultimate and most reliable manner of estimation of the quality of measurement results obtained by a given laboratory is the comparison of their results with those obtained in other laboratories. Bearing this in mind, laboratories for many years have participated in various interlaboratory comparisons, both on a national and international scale.

A major task in interlaboratory comparisons is the help offered to a laboratory in detecting all types of irregularities during a given analytical procedure that may affect the reliability of the obtained results. In other words, it is a system of mutual aid where a participant obtains information whether and how they should modify the applied measurement procedure to increase the reliability of the obtained results.

High marks/grades obtained in interlaboratory proficiency studies indicate a high quality of analyses performed by the participating laboratory. The test of the interlaboratory proficiency is used to estimate the reliability of determination results and is the basis for the validation of analytical procedures according to EN 17025, and enables issuance of opinions on organizational procedures. It is hence obvious that laboratories that do not participate in these comparisons are deemed unreliable.

However, while interpreting the results of the interlaboratory studies, one should remember that

- Participation in interlaboratory studies must not serve as a substitute for routine intralaboratory control of the results' quality.
- The results of the interlaboratory studies enable detection and definition of current problems in a given laboratory, and not those that may occur.
- A successful outcome in interlaboratory studies obtained during the determination of a given analyte or a group of analytes may not be automatically related to another analyte or group of analytes; the same applies to an analytical method.

To sum up, the major task of interlaboratory studies is to obtain an explicit answer to this question: Are the measurement results obtained in a given laboratory as good as we think they are? (See http://www.hn-proficiency.com.)

REFERENCES

1. Proficiency testing by interlaboratory comparisons, Part 1: Development and operation of proficiency testing schemes, ISO/IEC Guide 43-1, 1997.
2. Proficiency testing by interlaboratory comparisons, Part 2: Selection and use of proficiency testing schemes by laboratory accreditation bodies, ISO/IEC Guide 43-1, 1997.

3. Thompson M., and Ellison S.L.R., Fitness for purpose – the integrating theme of the revised harmonized protocol for proficiency testing in analytical chemistry laboratories, *Accred. Qual. Assur.*, 11, 373–378, 2006.

4. Juniper I.R., Quality issues in proficiency testing, *Accred Qual. Assur.*, 4, 336–341, 1999.

5. Konieczka P., The role of and place of method validation in the quality assurance and quality control (QA/QC) System, *Crit. Rev. Anal. Chem.*, 37, 173–190, 2007.

6. Analytical Methods Committee, Proficiency testing of analytical laboratories: Organization and statistical assessment, *Analyst*, 117, 97–104, 1992.

7. Vander Heyden Y., and Smeyers-Verbeke J., Set-up and evaluation of interlaboratory studies, *J. Chromatogr. A.*, 1158, 158–167, 2007.

8. Tholen D.W., Statistical treatment of proficiency testing data, *Accred. Qual. Assur.*, 3, 362–366, 1998.

9. Davies P.L., Statistical evaluation of interlaboratory tests, *Fresenius Z. Anal. Chem.*, 331, 513–519, 1988.

10. Eurachem Guide on Selection, Use and Interpretation of PT Schemes, Edition 1.0, 2000.

11. Linsinger T.P.J., Kandel W., Krska R., and Grasserbauer M., The influence of different evaluation techniques on the results of interlaboratory comparisons, *Accred. Qual. Assur.*, 3, 322–327, 1998.

8 Method Validation

8.1 INTRODUCTION

Considerations concerning the determination of validation parameters should begin with an explanation and description of the nature of an analytical measurement. The key interests of analysts worldwide are the signals following and resulting from a conducted measurement. The goal of an analyst's work is to obtain analytical information about an investigated object based on a received output signal, a result of a suitable measurement method. This signal reveals information about the investigated sample. The analyst's role is to "decode" the obtained signal and do it in a manner such that the obtained information is as reliable as possible [1]. A tool that decodes information is an analytical process, including analytical methods applied in the process.

Each signal is characterized by a particular quantity. In some measurements, a signal may also be assigned a position (location). Validation parameters are determined based on analysis of the obtained signal values, and one should be aware of this in the validation of any analytical method.

Validation of an analytical method includes testing of its important characteristics. The final aim is to be certain that the analysis process is reliable and precise, remains under total control of the operator, and leads to reliable results.

First of all, validation allows definition of a given analytical method. Using the determined parameters, in the validation process there exists the possibility of estimating the usefulness (range of use) for a given method and then choosing the optimal method.

As previously stated, for the measurement results to be traceable and have an uncertainty value provided, they must be obtained using an analytical method that is subjected to a prior validation process.

Most often, a validation study is carried out when [2,3]:

- Analytical method is being developed.
- Tests for the extension of the applicability of a known analytical method are being conducted, for example, determinations of a given analyte, but in samples characterized by a different matrix composition.
- Quality control of the applied method showed variability of its parameters over time.
- A given analytical method has to be used in another laboratory (different from the one in which it has already been subjected to the validation process), or different instruments have been used or determinations have been performed by another analyst.
- A comparison of a new analytical method with another known reference method is being performed.

The parameter range, the determination of which should underlie the validation process for a given analytical method, depends on the following factors [4]:

- The character of an analytical study to be carried out using a given analytical method (qualitative or quantitative analysis, analysis of a single sample, or a routine analytical investigation)
- Requirements for a given analytical method
- Time and costs, which need to be spent in the validation process

The parameters considered necessary for the validation of different types of analytical procedures are presented in Table 8.1 [2,5].

The more parameters included in the validation process, the more time one should spend on the process. In addition, the more restrictive the assumptions for the limit values (expected) of the respective parameters, the more often one should test, calibrate, or "revalidate" a given analytical method. It is not always necessary to conduct a full analytical method validation. Therefore, one should determine which parameters should be included in the process.

Table 8.2 contains the parameters which, according to the recommendations of the International Conference on Harmonization (ICH) [6,7] and the United States Pharmacopeia (USP) [8], should be included in the validation process.

TABLE 8.1
Parameters Whose Determination Is Necessary for Different Types of Analytical Procedures

| | | Impurity Test | | |
| | | | |
Parameter	Qualitative Analysis	Limit Impurity Test	Quantitative Impurity Test	Assay Test
Precision	− [a]	−	+	+
Correctness	−	− [a]	+	+
Specificity	+	+	+	+
Limit of Detection	− [a]	+	−	−
Limit of Quantitation	− [a]	−	+	−
Linearity	− [a]	−	+	+
Measuring range	− [a]	− [a]	+	+
Ruggedness	+	+	+	+

Source: Huber, L., Validation of analytical methods, http://www.chem.agilent.com/Library/primers /Public/5990-5140EN.pdf, 2010 (access date July 20, 2017); Traverniers, I., De Loose, M., and Van Bockstaele, E., *Trends Anal. Chem.*, 23, 535–552, 2004.

[a] It might be determined.

TABLE 8.2
List of Analytical Procedure Parameters that Should Be Validated According to the Recommendations of ICH and USP

Parameter	ICH	USP
Precision		
Repeatability	+	+
Intermediate precision	+	
Reproducibility	+	
Accuracy	+	+
Limit of detection	+	+
Limit of quantification	+	+
Specificity/selectivity	+	+
Linearity	+	+
Measuring range	+	+
Robustness		+
Ruggedness		+

Source: International Conference on Harmonization (ICH) of Technical Requirements for the Registration of Pharmaceuticals for Human Use, Text on Validation of Analytical Procedures, ICH-Q2A, Geneva, 1994; International Conference on Harmonization (ICH) of Technical Requirements for the Registration of Pharmaceuticals for Human Use, Validation of Analytical Procedures: Metrology, ICH-Q2B, Geneva, 1996; United States Pharmacopeial Convention, United States Pharmacopeia 23, US Rockville, 1995.

Apart from determining validation parameters, before commencing validation one should determine the basic features of an analytical method, namely [2]:

- Type of the determined component (analyte)
- Analyte concentration
- Concentration range
- Type of matrix and its composition
- Presence of interferents
- Existence of top-down regulations and requirements for the examined analytical method
- Type of the expected information (quantitative or qualitative analysis)
- Required limits of detection and quantitation
- Expected and required precision and accuracy of the entire method
- Required robustness of the method
- Required instruments; whether the determinations using a given method have to be carried out using a strictly defined measuring instrument or instruments of a similar type
- Possibility of using a method already validated in another laboratory(ies)

A validation process may be conducted in any order, however it seems most logical to proceed in the following manner [2,4]:

- Determine the selectivity in the analysis of standard solution samples (optimization of the separation conditions and determination of analytes present in the standard solution samples).
- Determine the linearity, limits of detection and quantitation, and the measuring range.
- Determine the repeatability (short-term precision), for example, based on deviations of the obtained retention times or chromatographic peak areas.
- Determine the intermediate precision.
- Determine the selectivity based on results obtained in the analyses of real samples.
- Determine the accuracy/trueness based on the analysis of reference material samples containing an analyte at different concentration levels.
- Determine the robustness of a method, for example, based on the results obtained in interlaboratory comparisons.

The validation process requires the use of various tools such as [9]

- Blank samples (including so-called reagent blanks)
- Standard solutions (calibration solutions, test samples)
- Samples with a known quantity of added analyte (spiked with the analyte)
- (Certified) reference materials
- Repetitions
- Statistical processing of the results

In this work, we need to stress that the method can be subjected to the validation process only when a suitable optimization study has been conducted.

The process of analytical method validation should be completed with the final report, which includes all information concerning the analytical method.

Validation parameter definitions and the manner of their determination are described below.

8.2   CHARACTERIZATION OF VALIDATION PARAMETERS

8.2.1   SELECTIVITY

Usually the first determined validation parameter is selectivity. Using basic logic, before one commences determination of the properties of an analyte based on measurement of the obtained analytical signal, one should make sure that a given signal is due only to the occurrence of an analyte in an investigated sample.

A quite frequent problem is the interchangeable use of the terms *selectivity* and *specificity*, although they differ in their essential meaning.

According to the International Union of Pure and Applied Chemistry (IUPAC) nomenclature [10], selectivity is defined as "the extent to which it can determine particular analyte(s) in a complex mixture without interference from other components in the mixture." Specificity is described by the IUPAC as the "highest selectivity" and is not recommended for use.

Selectivity is thus the ability of a method to differentiate the examined analyte from other substances. This characteristic is mostly a function of the described measurement technique, but can fluctuate depending on the class or group of compounds to which the analyte belongs, or the sample matrix. A specific method is one which shows the highest selectivity.

Selectivity can be defined as follows [11]: "The ability of an analytical process to receive signals whose size depends almost entirely on the concentration of the examined analyte present in the sample."

One can also propose a practical definition [9]: "Selectivity is the potential for an accurate and precise determination of the occurrence and/or concentrations of an analyte or groups of analytes in the presence of other components in a real sample under given measurement conditions."

Selectivity is therefore one of the main parameters characterizing and describing an analytical method, especially a trace analysis [12].

From a practical point of view, an analytical measurement is selective when it is possible to differentiate measurement signals and assign to them respective properties for a given analyte. This undoubtedly depends on the parameters of the obtained signal. If the signal is characterized only by its intensity, one should prove that its size depends only on the investigated properties of a given object. For example, if the mass of a sample is being determined using an analytical balance, then an analyst must be certain that the measured value is due to the real mass of a sample and not, for example, refuse on the balance's tray. This example shows that problems related to selectivity are also linked with direct measurements.

A different situation is observed concerning selectivity when signals are characterized by an additional parameter—position (place). Such a situation takes place in chromatography, for example, where retention time additionally characterizes the output signal and assigns it to a specific analyte. In such a case, it becomes necessary to determine the smallest differences between the positions for each analyte, for which the distinction between the obtained signals is possible.

The requirement of selectivity for a measurement process depends first of all on the composition of an analyzed sample [11]. Selectivity is more difficult to obtain:

- The more unknown the sample composition is
- The more complex the sample's matrix composition is
- The more similar the properties of the matrix components
- The greater the number of analytes
- The smaller the analyte concentration
- The greater the resemblance between analytes

An increase in the selectivity of an analysis may be obtained by

- The use of selective analytical methods
- Elimination of the influence of interferents by removing or concealing them
- Isolation of the analyte from the matrix

Depending on the type of analytical technique, the various ways of expressing selectivity are different.

8.2.2 LINEARITY AND CALIBRATION

When an investigated property is certain to be associated with a given signal, one should determine the dependence between these quantities. A linear dependence most frequently occurs in analytical chemistry. The vast majority of analytical measurements use the calibration step, when the output signals are assigned to corresponding analyte concentrations [13]. To determine the functional dependency associating the output signal with analyte concentration, the linear regression method is commonly used. It is also applied in the determination of some validation parameters, such as

- Linearity
- Trueness (based on the value of biases)
- Limits of detection and quantitation

It is also widely used in the calibration of measuring instruments.

Linearity is defined as an interval in the measurement range of an analytical method in which an output signal correlates linearly with the determined analyte concentration.

The most frequent manner of determining linearity is by using a graph of measuring instrument calibration. To this end, measurements of standard solution samples are conducted on at least six levels of concentrations (most often three parallel measurements for each level). Naturally, the selection of analyte concentrations in standard solution samples should be such that their range should include the expected analyte concentration in an investigated sample (the concentration range usually covers values from 50 percent to 150 percent in relation with the expected results of an analysis) [14]. Then, using the linear regression method, one determines the regression parameters.

According to some recommendations [15], it is sufficient to calculate the coefficient of regression. Then, if this value is at levels equal to at least 0.999, we may talk about the linearity of the method within the range of concentrations for which standard solutions were prepared to determine the calibration graph.

Unfortunately, this manner of documenting linearity does not always lead to correct conclusions. It can happen that the high value obtained for the coefficient of regression r (or the coefficient of determination r^2) does not necessarily prove the linearity of a method.

The coefficient of regression may be used to infer the linearity of an analytical method only when standard solutions, based on which the calibration curve is determined, fulfill the following requirements [14,16,17]:

- They include the expected analyte concentration in the investigated sample(s) within their own range of concentrations.
- They include no more than three orders of magnitude of analyte concentrations within their own range.
- They evenly "cover" the whole range of concentrations.

In addition, it is very important to determine a suitable dependence and the "visual" analysis of the obtained graph.

Because of the ambiguity in usage of the coefficient r as a measure of linearity, additional methods for proving linearity have been proposed.

In addition, the significance of the calibration graph coefficients needs to be determined. The slope should differ statistically and significantly from 0, and in the case of an intercept, its value should not differ in a statistically significant way from 0. To ascertain this, one should calculate the values of Student's t (Section 1.8.9).

Another approach is to draw a so-called graph of constant response described by the following dependence [2]:

$$\frac{y}{x} = f(x) \tag{8.1}$$

where
  y: Signal of a measuring instrument
  x: Analyte concentration in a standard sample corresponding to a given signal

When the range of concentrations is sufficiently large (including three or more orders of magnitude), the concentrations may be marked on the graph in a logarithmical scale. On such a graph, the sustained response is marked (calculated usually as an arithmetical mean of individual values y/x) in the form of a line parallel to the X-axis, along with the admissible deviations from this value (most often ±5 percent). Values (points) lying outside the determined range correspond to analyte concentrations that lie outside the linear range of the measuring instrument.

Naturally, this process can only be used when an intercept of the determined simple dependence $y = f(x)$ does not differ in a statistically significant manner from zero, which is not always the case.

In some studies, one can find unambiguous and categorical statements that the value of coefficient r cannot serve to determine the degree of dependence between variables, and should be replaced by another statistical tool or specific tests for proving linearity [18]. One of the recommended tools is variance analysis. One can also use other methods and statistical tools such as [19–22]

- Test of adequacy
- Mandel's test
- Quality factor
- Student's t test (Section 1.8.9)

When proving linearity is based on analysis results of the standard solution series with the simultaneous drawing of a calibration graph, it is logical to prove to what extent the calibration curve reflects the signals for standard solution samples. One can ascertain this through the calculation of relative errors for each concentration, with the reference value being the analyte concentration in the standard sample, and the experimental value being that calculated from the equation of a straight calibration line [23].

Linearity by no means signifies that within the entire range of concentrations the function describing the dependence of the output signal on the analyte concentration assumes one form (the same calibration curve coefficients). Linearity is a characteristic showing the linear dependence of a signal on the determined quantity, and can be described for a given range by several equations depending on the level of analyte concentrations [24,25].

It is also necessary to explain the difference between correlation and regression. Correlation describes the degree of connection between two variables, and regression describes the manner of their dependence [18].

Example 8.1

Problem: Draw the calibration curve based on the results of the analyte concentration determination results in six standard solution samples (three independent measurements per each of the solutions). Calculate the regression parameters of the calibration curve.

Make an appropriate graph.

Data: Results:

	Data	
	x	y
1	2	1.12
2	2	1.20
3	2	1.08
4	4	2.11
5	4	2.32
6	4	2.23
7	6	3.33
8	6	3.54
9	6	3.41
10	8	4.12
11	8	4.32
12	8	4.44
13	10	5.67
14	10	5.76
15	10	5.51
16	12	6.97
17	12	6.78
18	12	6.66

Solution:

n	18
Slope—*b*	0.5642
Intercept—*a*	−0.029
Residual standard deviation—SD_{xy}	0.14
Standard deviation of the slope—SD_b	0.0099
Standard deviation of the intercept—SD_a	0.077
Regression coefficient—*r*	0.9976

Graph:

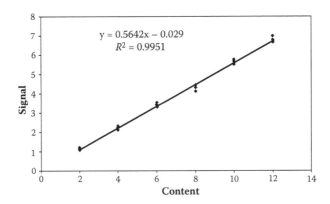

$$y = 0.5642x - 0.029$$
$$R^2 = 0.9951$$

Excel file: exampl_valid01.xls

Example 8.2

Problem: Using the data from the Example 8.1, examine the significance of the differences in the slope and the intercept of a calibration line and the value 0. Apply Student's *t* test.

Calculations should be performed for the significance level $\alpha = 0.05$.

Data: Results:

	Data	
	x	*y*
1	2	1.12
2	2	1.20
3	2	1.08
4	4	2.11
5	4	2.32

6	4	2.23
7	6	3.33
8	6	3.54
9	6	3.41
10	8	4.12
11	8	4.32
12	8	4.44
13	10	5.67
14	10	5.76
15	10	5.51
16	12	6.97
17	12	6.78
18	12	6.66

Solution:

n	18
Slope—b	0.5642
Intercept—a	−0.029
Residual standard deviation—SD_{xy}	0.14
Standard deviation of the slope—SD_b	0.0099
Standard deviation of the intercept—SD_a	0.077
Regression coefficient—r	0.9976
t_b	57.062
t_a	0.378
t_{crit}	2.120

Conclusions:
    Statistically significant difference between the slope and 0.
    No statistically significant difference between the intercept and 0.
Excel file: exampl_valid02.xls

Example 8.3

Problem: Using the data from the Example 8.1, draw a graph of sustained response, marking the lines of the interval for the values deviating ±5 percent from the mean.
Data: Results:

	Data	
	x	y
1	2	1.12
2	2	1.20

3	2	1.08
4	4	2.11
5	4	2.32
6	4	2.23
7	6	3.33
8	6	3.54
9	6	3.41
10	8	4.12
11	8	4.32
12	8	4.44
13	10	5.67
14	10	5.76
15	10	5.51
16	12	6.97
17	12	6.78
18	12	6.66

Solution:

	y/x
1	0.56
2	0.60
3	0.54
4	0.53
5	0.58
6	0.56
7	0.56
8	0.59
9	0.57
10	0.52
11	0.54
12	0.56
13	0.57
14	0.58
15	0.55
16	0.58
17	0.57
18	0.56

x_m	$x_m - \text{interval}\%$	$x_m + \text{interval}\%$
0.56	0.53	0.59

Graph:

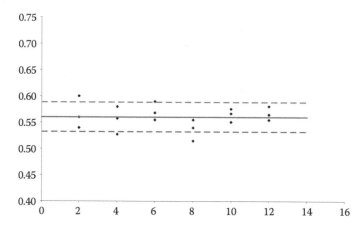

Excel file: exampl_valid03.xls

Example 8.4

Problem: Using the data from the Example 8.1, calculate the values of the relative errors for individual values x, assuming the reference value to be x and the experimental value to be the value calculated from the calibration curve equation.

Assume an appropriate limit for the relative error and draw conclusions.

Data: Results:

	Data	
	x	*y*
1	2	1.12
2	2	1.20
3	2	1.08
4	4	2.11
5	4	2.32
6	4	2.23
7	6	3.33
8	6	3.54
9	6	3.41
10	8	4.12
11	8	4.32
12	8	4.44
13	10	5.67
14	10	5.76
15	10	5.51
16	12	6.97
17	12	6.78
18	12	6.66

	Relative error—ε,%	5.00

Solution:

	Number of results—n	18
	Slope—b	0.5642
	Intercept—a	−0.029

	ε,%	Conclusion
1	1.83%	OK
2	8.92%	!!!
3	−1.72%	OK
4	−5.22%	!!!
5	4.08%	OK
6	0.10%	OK
7	−0.78%	OK
8	5.43%	!!!
9	1.59%	OK
10	−8.08%	!!!
11	−3.65%	OK
12	−0.99%	OK
13	1.01%	OK
14	2.60%	OK
15	−1.83%	OK
16	3.37%	OK
17	0.56%	OK
18	−1.21%	OK

Note: If the relative error is greater then assigned value (it the example 5%) it is signed as !!!!

Excel file: exampl_valid04.xls

The term *linearity* is highly connected with calibration. The most commonly used method for calibration is the calibration curve method. There are two main groups of calibration methods:

- External—Standard samples are measured independently from real samples
- Internal—Standard is usually added to the real sample and analyzed together

The following belong to the first group of calibrations:

- Single point calibration—One standard solution is used and the content of analyte in real sample is calculated for used proportional dependence between signals for standard and real sample.
- Bracketing solution calibration—Two standards are used: One with lower and one with higher signals than in real samples.
- Multipoint calibration—Classical calibration curve method.

In the second group we highlight:

- Standard addition method—known quantities of analyte are added to an unknown, and the analyte concentration is determined from the increase in signal. Standard addition is often used when the sample is unknown or complex and when species other than the analyte affect the signal.
- Internal standard method—known amount of a compound, different from the analyte, is added to the unknown sample. Internal standards are used when the detector response varies slightly from run-to-run because of hard-to-control parameters.

8.2.3 LIMIT OF DETECTION AND LIMIT OF QUANTITATION

The next validation parameters that need to be determined are the *LOD* and the *LOQ*. The values of these parameters are closely related to the magnitude of noises in the measurement system.

Signal-to-noise ratio (S/N) is a unidimensional quantity that describes the relationship of an analytical signal to the mean noise levels for a specific sample. The value of this parameter can serve to determine the influence of noise level on the relative measurement deviation. It can be calculated in different ways, but the most common method is the relationship of the arithmetical mean of the results in a measurement series for blank samples (or samples containing analyte in a very low level) to the standard deviation obtained for this series.

Limit of detection (*LOD*) is the lowest concentration (smallest quantity) of an analyte than can be detected with statistically significant certainty [26]; this value is *n*-times the noise level—it is most often three times as high.

Method detection limit (MDL) is the lowest concentration (smallest quantity) of an analyte that can be detected using a given analytical procedure.

Instrumental detection limit (e.g., detector) (IDL) is the lowest concentration (smallest quantity) of an analyte which can be detected (without quantitative determination) using a given measuring instrument.

Limit of quantitation (*LOQ*) is the quantity or the smallest concentration of a substance that can be determined using a given analytical procedure with an assumed accuracy, precision, and uncertainty. This value should be estimated using a suitable standard sample and should not be determined through extrapolation [27].

LOD and *LOQ* are parameters that play an unusually significant role in the validation of analytical procedures. Although the meaning of these parameters and their understanding do not raise questions, the determination of their values itself is sometimes problematic. This can be attributed to several reasons:

- A large number of definitions describing the notions of both the *LOD* and the *LOQ*
- Practical difficulties in univocally determining the basic parameter deciding the *LOD*—namely, the magnitude of the noise level in a given measuring instrument

The manner of determining an *LOD* depends on the following factors:

- Nature of the analytical method (the manual method and the method based on utilization of a suitable gauge as well)
- Characteristics of the applied instrumental technique
- Possibilities of obtaining (producing) so-called blank samples

Depending on these parameters, there exist several ways of determining (estimating) the *LOD*.

8.2.3.1 Visual Estimation

For a classical method (noninstrumental) for which it is not possible to determine the noise level of the applied measuring instrument, one estimates the *LOD* based on one's own experiment. Based on the results of sample analysis with the known analyte concentration (standard solutions), one estimates this concentration level at which detection is possible. This method can also be used for instrumental techniques.

8.2.3.2 Calculation of *LOD* Based on the Numerical Value of the S/N Ratio

When calculating the *LOD*, one uses the determined S/N ratio for the investigated analytical procedure [2]. This method can be applied only when it is possible to obtain the baseline of noises, obtained when a blank sample is subjected to final determination.

In this instance, the simplest and most commonly applied way of calculating the *LOD* is to determine the S/N ratio for a blank sample (if it is possible) or for a sample with a very low analyte concentration, and then to directly apply the principle that *LOD* is three times the noise level for an applied analytical method.

In the case of chromatography, one can determine the *LOD* value using the obtained chromatogram for a blank sample. To this end, one describes the noise level—measuring range signal changes close to the retention time for an analyte on a chromatogram (one can assume the retention time range as $t_{R\,an} \pm$ of 0.5 min). This quantity is then multiplied by 3 and the obtained signal value is converted into a concentration.

8.2.3.3 Calculation of *LOD* Based on Determinations for Blank Samples

A more labor-consuming method, but one that is also metrologically more correct, is using a measurement for a series of blank samples. It involves 10 independent measurements for 10 independently prepared blank samples [28].

For the thusly obtained 10 results, one calculates the mean value and the standard deviation. *LOD* is equal to the mean value magnified by three times the standard deviation in this instance.

$$LOD = x_m + 3 \cdot SD \qquad (8.2)$$

where
  x_m: Mean value
  SD: Standard deviation

In practice, however, it is seldom possible to obtain a numerical value for the mean; it seems paradoxical to obtain a result for a value which by definition should be a submarginal quantitation. The method would only have some application when the analyte concentration was measurable for a blank sample, that is, the so-called background level is above the *LOD* for the applied detector (that is to say, the analyte concentration in a blank sample is at least equal to the *LOQ* for the applied detector).

Otherwise, it is possible to use the described method with a certain modification [29,30]—namely, 10 independent determinations are performed for samples in which the analyte concentration is close to the expected *LOD*; of course, such samples are prepared through a spike in the blank samples with quantifiable amounts of the analyte. The manner of conduct is then similar to the previously described one, with the one difference being that the *LOD* is calculated according to the following formula:

$$LOD = 0 + 3 \cdot SD \tag{8.3}$$

The modification is the preparation of n samples with analyte concentrations on a level close to the expected *LOD*. Of course, it would be most convenient to prepare standard solutions in which matrix compositions correspond to the matrix composition of real samples. One then performs an analysis on such prepared samples, receiving a series of n results for which one calculates the mean value and standard deviation. *LOD* is calculated using a dependence described by an equation for the number of degrees of freedom $f = n - 1$, where n is the number of independent samples and the accepted level of significance α:

$$LOD = t \cdot SD \tag{8.4}$$

where
   t: Parameter of Student's t test
   SD: Standard deviation

If the prepared standard solution samples are subjected to analysis using a given analytical procedure, then the determined *LOD* is also the MDL. If determinations are instead performed directly on the prepared standard solution samples, then IDL are determined in this manner.

8.2.3.4   Graphical Method

This method involves analyses of measurement series for three standard solution samples containing an analyte at three levels of concentration (close to the expected *LOD* for the samples). For each level of analyte concentration, one should perform at least six parallel determinations, and then for each series of measurements obtained in this way, calculate the standard deviations. A linear dependence is determined which associates the calculated standard deviations with the respective concentrations:

$$SD = f(c) \tag{8.5}$$

Then one determines the absolute term SD_o after which one determines the *LOD* according to the following dependence:

$$LOD = 3 \cdot SD_o \tag{8.6}$$

8.2.3.5 Calculating *LOD* Based on the Standard Deviation of Signals and the Slope of the Calibration Curve

One of the most commonly applied analytical methods is that in which the final determination is based on the indirect measurement principle. In this case, it is indispensable to perform calibration that will influence the *LOD* [2,6,7].

In this case, *LOD* is calculated using the following dependence:

$$LOD = \frac{3.3 \cdot SD}{b} \qquad (8.7)$$

where
   b is the slope of calibration curve.

Standard deviation can be determined in three different ways:

- As a standard deviation of results obtained for the series of blank samples—SD_{bl}
- As a residual standard deviation of the calibration curve—SD_{xy}, described by the dependence (1.68)
- As a standard deviation of the intercept of the obtained calibration curve—SD_a, described by the dependence (1.67)

Of course, the limit of detection (for the analytical method or the applied detector) will be calculated depending on which parameters were used to calculate the standard deviation. Hence, if measurements are conducted based on analyses of blank samples subjected to the whole analytical procedure, the MDL is the determined quantity. When the *LOD* is calculated based on parameters of the determined calibration graph (residual standard deviation or standard deviation of the intercept), the calculated value is the *LOD* for the measuring instrument. It is also important to appropriately select concentrations of standard solutions to draw the calibration graph (it is known that the calibration graph has a straight-line range in a strictly specific interval of concentrations and that its plot most likely has different concentration level characteristics close to the *LOD*).

8.2.3.6 Calculation of *LOD* Based on a Given *LOQ*

LOQ is the lowest analyte concentration that can be determined with suitable precision and accuracy. One performs measurements for standard solutions (matrix standards) on at least five levels of concentrations [2,28]. For each solution, one performs six parallel measurements. For each level of concentrations, the coefficient of variation (CV) is calculated and the graph of the *f(c)* dependence is drawn. The required precision for the *LOQ* is determined (usually = 10 percent), and for this value the concentration equal to the LQ is read on the graph. The LD is calculated as *LOD* = *LOQ*/3.

Figure 8.1 presents the construction of the graph and the calculation of the *LOQ* [28].

Table 8.3 compares all described methods of calculating the *LOD* along with their short characterizations [27].

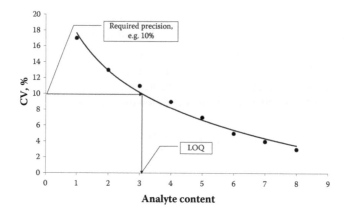

FIGURE 8.1 Construction of the graph and calculation of the limit of quantitation. (From EURACHEM Guide: The Fitness for Purpose of Analytical Methods, A Laboratory Guide to Method Validation and Related Topics, Second Internet Edition, 2014, https://www.eurachem.org/images/stories/Guides/pdf/MV_guide_2nd_ed_EN.pdf.)

8.2.3.7 Testing the Correctness of the Determined *LOD*

Many of the aforementioned ways of calculating the *LOD* are based on the determination of analyte concentration in the prepared standard solution samples. The solutions should be characterized, while calculating the *LOD* of an analytical procedure, with two basic features:

- Matrix composition should be as close to the matrix composition of real samples as possible
- Analyte concentration should be on a level close to the expected *LOD*

It is known that the standard deviation for the set of measurement results determining the analyte concentrations in standard solution samples strictly depends on the concentration levels of a determined component. It can happen that the concentrations in standard samples are considerably higher than the calculated *LOD*. To check the calculated *LOD*, one should fulfill the following conditions [29]:

$$10 \cdot LOD > c_{min} \tag{8.8}$$

$$LOD < c_{min} \tag{8.9}$$

where
c_{min} is the analyte concentration in a standard solution sample with the lowest concentration.

TABLE 8.3
Methods for Determining Detection Limits: Requirements, Disadvantages, and Advantages

Methods for Calculating *LOD*	Requirements	Disadvantages/Advantages
Visual check	Sample with known analyte content (standard solution or matrix standard)	Quick method Estimation Mostly used in case of classical analysis (noninstrumental) Requires vast analytical experience
Calculations based on the S/N ratio	Sample with known analyte content (standard solution or matrix standard)	Quick method Used only for measuring equipment It is possible to determine the S/N ratio
Calculations based on the measurements for sample blanks	Series of blanks or samples with known analyte content (standard solution or matrix standard)	Labor- and time-consuming method that does not consider the influence of calibration on *LOD* Probability is used for estimating *LOD*
Calculations based on graphical method	Series of standard samples at three concentration levels, at least six measurements for each standard sample	Relatively quick method It includes the influence of calibration procedure on *LOD* value
Calculations based on standard deviations of signals and slope of calibration curve	Series of blanks or samples with known analyte content (standard solution or matrix standard) Standard solutions for calibration curve preparation	Labor- and time-consuming method It includes the influence of calibration procedure on *LOD* value Method "motivated" by metrology
Calculations based on limit of quantifica-tion, *LOQ*	Series of standard solutions Assumed relative standard deviation for *LOQ*	Indirect method *LOD* calculated based on the determined *LOQ* *LOD* value (*LOQ*) depends on the assumed measurement precision

Source: Konieczka, P., Sposoby wyznaczania granicy wykrywalności i oznaczalności, *Chem. Inż. Ekol.*, 10, 639–654, 2003 (in Polish).

If condition (8.8) is not fulfilled, it will signify that the concentration in the prepared standard samples is too high. One should then calculate the *LOD* for newly prepared standard solutions with a lower analyte concentration. Inversely, when the condition (8.9) is not fulfilled, the analyte concentration in the prepared standard samples is too low. In this case, one should remeasure and recalculate using standard solutions in which the analyte occurs in higher concentration levels.

In order to test the trueness of the calculated *LOD*, one can also estimate the S/N ratio based on the following dependence [29]:

$$S/N = \frac{x_m}{SD} \tag{8.10}$$

According to the definition of *LOD*, the numerical value of this ratio should be between 3 and 10. When it is higher, the determined *LOD* is greater than the numerical value and one should conduct remeasurement for lower concentrations of the analyte in standard solution samples.

One should also pay attention to the recovery of the analytical method in measurements conducted for standard solutions. Recovery can be calculated using the following dependence [29]:

$$\%R = \frac{x_m}{c}[\%] \tag{8.11}$$

where

%R: Recovery of an analyte for a given analytical procedure

A recovery being too low results in an undervaluation of the calculated *LOD*.

As previously stated, the described methods of testing the correctness of the calculated *LOD* can only be applied when the measurements are performed using prepared standard solutions.

The described ways of determining the *LOD* or quantitation permit the determination of both the MDL and the IDL.

The choice of a suitable means for determining the *LOD* depends on the purpose of the limit and the requirements of a given analytical method. For validation of an analytical method, it is recommended to use a way the assumptions of which are based on chemical metrology; the value of the determined *LOD* is associated with statistical parameters such as

- Level of probability
- Number of degrees of freedom

For individual measurements, it is recommended to apply a less time-consuming method.

It must be stated that the determined *LOD* should always be given the description and parameters of the method applied in its calculation.

The determined limits of detection and quantitation also show the quality of measurements conducted using a given analytical method [31,32].

Determining limits of detection and quantitation allows the unequivocal determination and presentation of results in the proximity of these values. A correct method

---

TABLE 8.4

Correct Method for Recording a Determination Result

Result, x	Recording of Result
$x < LOD$	Not determined
$LOD \leq x < LOQ$	Not quantified
$x \geq LOQ$	Value of concentration

---

for recording a determination result depending on the quantity of an analytical signal is presented in Table 8.4.

It has to be stressed that both LOD and LOQ are the parameters which are estimated. It means that its presentation should have a maximum of two significant digits.

Example 8.5

Problem: Using the given analyte concentration determinations for blank samples, estimate LOD and LOQ for the validated analytical method.

Using the calculated S/N ratio, examine the correctness of the determined LOD.

Data: Results, ng/g:

	Data
1	0.155
2	0.132
3	0.143
4	0.121
5	0.145
6	0.113
7	0.137

Solution:

x_m	0.135	ng/g
SD	0.014	ng/g
LOD	0.18	ng/g
LOQ	0.54	ng/g

$$LOD = x_m + 3 \cdot SD$$

$$S/N = \frac{x_m}{SD}$$

S/N	9

Conclusion: *S/N* ratio is in the range 3 ÷ 10; calculated *LOD* is correct.
Excel file: exampl_valid05.xls

Example 8.6

Problem: During the measurements performed on blank samples it was noticed that the obtained values of signals cannot be measured. Hence, the standard solutions were made with concentrations near the expected *LOD*, and based on the measurements for these solutions the estimation was made for *LOD* and *LOQ*.

Check the correctness of the *LOD* determination through the comparison with the standard solution concentration.

Data: Results, ng/g:

	Data
1	0.235
2	0.253
3	0.258
4	0.254
5	0.244
6	0.258
c	**0.250**

Solution:

SD	0.0091	ng/g
LOD	0.027	ng/g
LOQ	0.082	ng/g

$$LOD = 0 + 3 \cdot SD$$

$$10 \cdot LOD > c_{min}$$

$$LOD < c_{min}$$

Conclusion: Calculated *LOD* is lower than the standard solution concentration used for its determination and 10 times *LOD* is higher than the standard solution concentration; calculated *LOD* is correct.
Excel file: exampl_valid06.xls

Example 8.7

Problem: Using the given data for determinations of the analyte concentrations for blank samples, estimate the *LOD* and *LOQ* of the validated analytical method, using Student's *t* test.
   Using the data calculated *S/N* ratio, check correctness of the determined *LOD*.
Data: Results, mg/dm³:

	Data
1	8.8
2	7.6
3	9.2
4	9.5
5	6.8
6	7.4
7	9.6

α	0.05

Solution:

x_m	8.41	mg/dm³
SD	1.13	mg/dm³
t	2.447	
LOD	2.8	mg/dm³
LOQ	8.3	mg/dm³

$$LOD = t \cdot SD$$

$$S/N = \frac{x_m}{SD}$$

S/N	7

Conclusion: *S/N* ratio is in the range 3 ÷ 10; calculated *LOD* is correct.
Excel file: exampl_valid07.xls

Example 8.8

Problem: Using the given analyte concentration determinations for standard solution samples, estimate the *LOD* and *LOQ* using a graphical method. Draw an appropriate graph. Present the *LOD* in units of the analyte concentration in standard solutions applied for *LOD* estimation.

In addition, check correctness of the *LOD* determination through a comparison with the standard solution with the lowest concentration.

Data: Results:

	Concentration, ppm		
	0.11	0.15	0.23
	Signals		
1	101	198	298
2	144	177	237
3	124	132	222
4	174	156	257
5	102	205	243
6	111	193	313
7	121	135	235

Solution:

Concentration	Signal	SD (signal)
0.11	125.3	26.1
0.15	170.9	30.1
0.23	257.9	34.4
SD_o	19.2	signal
LOD	0.013	ppm
LOQ	0.040	ppm

Graph:

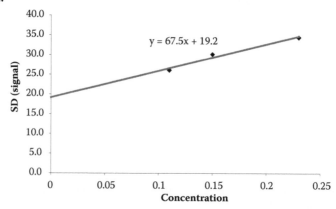

y = 67.5x + 19.2

Graph_calibration:

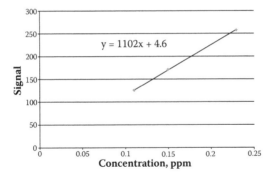

$$10 \cdot LOD > c_{min}$$

$$LOD < c_{min}$$

Conclusion: Calculated *LOD* is lower than the lower concentrated standard solution used for its determination and 10 times *LOD* is higher than the lower concentrated standard solution; calculated *LOD* is correct.
Excel file: exampl_valid08.xls

Example 8.9

Problem: Using the data from the Example 8.8, estimate the *LOD* and *LOQ* via the method using parameters of the calibration curve.
Present the value of *LOD* in the units of standard solution concentration, applied in *LOD* estimation.
Also check the correctness of *LOD* determination, comparing the calculated value with the value of the analyte concentration in the standard solution with the lowest concentration.
Data: Results:

	Concentration, ppm	Signal
1	0.11	101
2	0.11	144
3	0.11	124
4	0.11	174
5	0.11	102
6	0.11	111
7	0.11	121
8	0.15	198
9	0.15	177
10	0.15	132
11	0.15	156

12	0.15	205
13	0.15	193
14	0.15	135
15	0.23	298
16	0.23	237
17	0.23	222
18	0.23	257
19	0.23	243
20	0.23	313
21	0.23	235

Solution:

Number of results—n	21
Slope—b	1102
Intercept—a	4.6
Residual standard deviation—SD_{xy}	29.6
Standard deviation—SD_b	129
Standard deviation—SD_a	22.1
Regression coefficient—r	0.8901

LOD ($SD_{x,y}$)	0.089	ppm
LOD (SD_a)	0.066	ppm
LOD (mean)	0.077	ppm

$$LOD = \frac{3.3 \cdot SD}{b}$$

Graph:

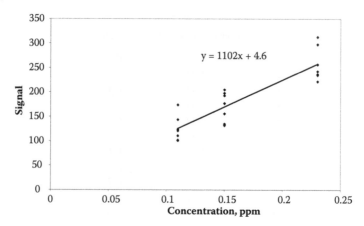

$$10 \cdot LOD > c_{min}$$

$$LOD < c_{min}$$

Conclusion: The calculated *LOD* is lower than the lower concentrated standard solution used for its determination and 10 times *LOD* is higher than the lower concentrated standard solution; calculated *LOD* is correct.
Excel file: exampl_valid09.xls

Example 8.10

Problem: Using the analyte concentration determinations for standard solution samples, estimate the *LOD* and *LOQ* by a method using the parameters of the calibration curve.

Present the values of *LOD* in units of standard solution concentrations applied for *LOD* determination.

Also check the correctness of *LOD* determination comparing the calculated value with the analyte concentration in a standard solution with the lowest concentration.

Data: Results:

	Concentration, ppm	Signal
1	1.2	1460
2	1.2	1725
3	1.2	1150
4	1.2	1025
5	1.2	1825
6	1.2	1310
7	2.5	1950
8	2.5	1630
9	2.5	2200
10	2.5	1650
11	2.5	2000
12	2.5	1980
13	3.3	2900
14	3.3	3200
15	3.3	3245
16	3.3	2850
17	3.3	3500
18	3.3	3890

Solution:

Number of results—n	18
Slope—b	831
Intercept—a	254
Residual standard deviation—SD_{xy}	447
Standard deviation—SD_b	122
Standard deviation—SD_a	303
Regression coefficient—r	0.8627

$LOD\ (SD_{xy})$	1.8	ppm
$LOD\ (SD_a)$	1.2	ppm
$LOD\ (mean)$	1.5	ppm

$$LOD = \frac{3.3 \cdot SD}{b}$$

Graph:

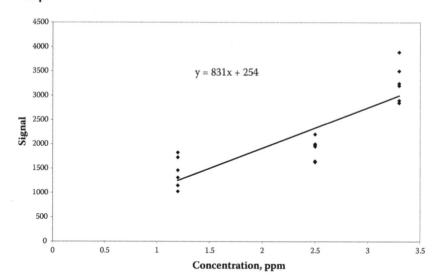

y = 831x + 254

$$10 \cdot LOD > c_{min}$$

$$LOD < c_{min}$$

Conclusion: Because the concentration of a solution with the lowest concentration is lower than the calculated *LOD*, standard solutions with a higher concentration were made and new calculations were made for the new series of data (without measurements for the solution with the lowest concentration).

Data (2): Results:

	Concentration, ppm	Signal
1	2.5	1950
2	2.5	1630
3	2.5	2200
4	2.5	1650
5	2.5	2000
6	2.5	1980
7	3.3	2900
8	3.3	3200
9	3.3	3245
10	3.3	2850
11	3.3	3500
12	3.3	3890
13	4.7	3640
14	4.7	4650
15	4.7	3860
16	4.7	4750
17	4.7	4450
18	4.7	4025

Solution (2):

Number of results—*n*	18
Slope—*b*	1016
Intercept—*a*	−425
Residual standard deviation—SD_{xy}	437
Standard deviation—SD_b	113
Standard deviation—SD_a	410
Regression coefficient—*r*	0.9132

LOD (SD_{xy})	1.4	ppm
LOD (SD_a)	1.3	ppm
LOD (mean)	1.4	ppm

Graph (2):

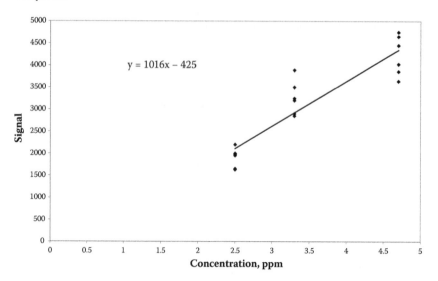

$y = 1016x - 425$

Conclusion (2): The calculated *LOD* is lower than the lower concentrated standard solution used for its determination and 10 times *LOD* is higher than the lower concentrated standard solution; calculated *LOD* is correct.

Excel file: exampl_valid10.xls

Example 8.11

Problem: Using the values of the analyte concentration determinations for standard solution samples, estimate the *LOQ* and then the *LOD* using a *LOQ* determination method based on the assumed value of determination precision. Assume the maximum value of the coefficient of variation to be *CV* = 5 percent.

 Draw an appropriate graph.

 Present the *LOD* in units of the standard solution concentration applied for *LOD* determinations.

Data: Results:

| | Concentration, ppm | | | | | |
	5.0	10.0	20.0	30.0	40.0	50.0
			Signals			
1	104	198	444	635	800	1000
2	144	177	450	650	810	990
3	124	232	470	660	805	995
4	124	200	400	620	825	1010
5	102	205	445	610	820	1005
6	111	193	450	625	840	1015
7	121	235	470	615	830	995

Solution:

Concentration, ppm	CV,%
5.0	12.2
10.0	10.2
20.0	5.24
30.0	2.95
40.0	1.75
50.0	0.898

LOQ	22	ppm
LOD	7.3	ppm

Graph:

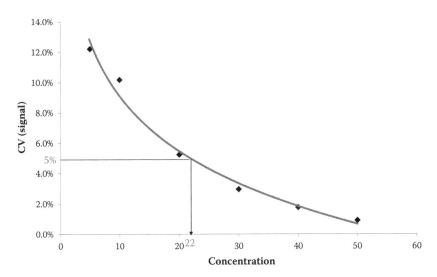

Excel file: exampl_valid11.xls

8.2.4 RANGE

Determination of linearity and the *LOQ* enables the determination of a measuring range for an analytical method. A *measuring range* is a range of values (analyte concentrations) in which the error of a measuring instrument is below the assumed value. In practice, it is described as an interval between the *LOQ* and the highest analyte concentration for which a measuring system shows an increase in the output signal.

Example 8.12

Problem: Determine the calibration curve based on analyte concentration determinations in eight standard solutions samples (five independent measurements for
each solution). Calculate regression parameters of the calibration curve.

Prepare an appropriate graph.

Using the determinations for standard solution samples for the three lowest
concentration levels, estimate the LOD and LOQ using a technique based on using
parameters of the calibration curve.

Present the LOD in units of standard solution concentration applied in LOD
estimation.

Also check the correctness of the LOD determination by comparing the calculated value with the analyte concentration in the standard solution with the lowest
concentration.

Present the measuring range of the analytical method.

Data: Results:

	Concentration, ppb	Signal
1	0.65	780
2	0.65	745
3	0.65	756
4	0.65	770
5	0.65	735
6	1.12	1420
7	1.12	1450
8	1.12	1425
9	1.12	1350
10	1.12	1411
11	2.44	3100
12	2.44	3005
13	2.44	3000
14	2.44	3100
15	2.44	3105
16	3.75	4700
17	3.75	4650
18	3.75	4850
19	3.75	4760
20	3.75	4690
21	5.25	6750
22	5.25	6800
23	5.25	7100
24	5.25	6690
25	5.25	6990
26	7.8	10100
27	7.8	10000
28	7.8	9900
29	7.8	10350

30	7.8	10150
31	10.4	13400
32	10.4	13200
33	10.4	13300
34	10.4	13000
35	10.4	12950
36	13.3	16600
37	13.3	16745
38	13.3	16600
39	13.3	16200
40	13.3	16500

Solution (calibration):

Number of results—n	40
Slope—b	1255.9
Intercept—a	59
Residual standard deviation—SD_{xy}	200
Standard deviation—SD_b	7.4
Standard deviation—SD_a	52
Regression coefficient—r	0.9993

Solution (LOD):

Number of results—n	15
Slope—b	1280
Intercept—a	−52
Residual standard deviation—SD_{xy}	45
Standard deviation—SD_b	15
Standard deviation—SD_a	24
Regression coefficient—r	0.9991

LOD (SDxy)	0.12	ppb
LOD (SDa)	0.063	ppb
LOD (mean)	0.089	ppb
LOQ	0.27	ppb
Range	0.27 ÷ 13.3	ppb

$$LOD = \frac{3.3 \cdot SD}{b}$$

Graph (calibration):

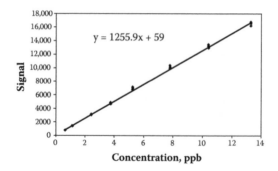

Graph (LOD):

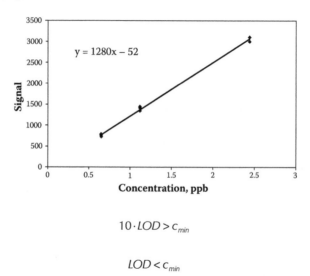

$$10 \cdot LOD > c_{min}$$

$$LOD < c_{min}$$

Conclusion: Calculated *LOD* is lower than the lower concentrated standard solution used for its determination and 10 times *LOD* is higher than the lower concentrated standard solution; the calculated *LOD* is correct.
Excel file: exampl_valid12.xls

8.2.5   SENSITIVITY

Sensitivity is a parameter that is not necessary in the validation of an analytical method. One can determine its value based simply on the parameters of the calibration curve. *Sensitivity* is the relationship of change in the output signal of a measuring instrument to the change in the analyte concentration that induces it. Thus, sensitivity shows the smallest difference in the analyte concentration that can be ascertained using a specific method (it is a slope of a calibration graph: signal in the concentration function).

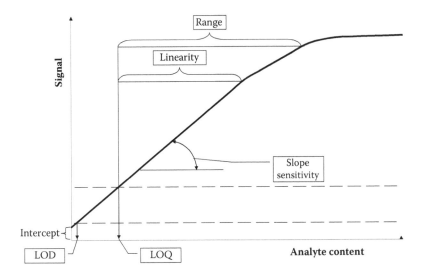

FIGURE 8.2 Interpretation of linearity, measuring range, limit of detection, limit of quantitation, and sensitivity. (From EURACHEM Guide: The Fitness for Purpose of Analytical Methods, A Laboratory Guide to Method Validation and Related Topics, Second Internet Edition, 2014, https://www.eurachem.org/images/stories/Guides/pdf/MV _guide_2nd_ed_EN.pdf.)

As a recapitulation, Figure 8.2 presents the interpretation of linearity, measuring range, *LOD*, *LOQ*, and sensitivity [28].

8.2.6 PRECISION

Each of the parameters below is determined based on the calculated standard deviation for the series of measurements, and therefore the manner of conduct in their determination will be described together.

Repeatability, intermediate precision, and reproducibility can be determined based on the determined standard deviation, relative standard deviation, or the so-called coefficient of variation.

Precision is the closeness of agreement between indications or measured quantity values obtained by replicate measurements on the same or similar objects under specified conditions [26]. It is associated with random errors and is a measure of dispersion or scattering around the mean value, usually expressed by a standard deviation.

Repeatability is the measurement precision under a set of repeatability conditions of measurement [26]. The precision of results obtained under the same measurement conditions (a given laboratory, analyst, measuring instrument, reagents, etc.). It is usually expressed by a repeatability standard deviation, variance, relative standard deviation, or coefficient of variation.

Intermediate precision is the precision of results obtained in a given laboratory over a long-term process of measuring. Intermediate precision is a more general notion (due to the possibility of changes in the greater number of determination parameters) compared to repeatability.

Reproducibility is the precision of results obtained by different analysts in different laboratories using a given measurement method. For or during instead of in determining repeatability, it is recommended for an analysis to be conducted on samples characterized with different analyte concentrations and differing in matrix composition.

According to recommendations by the ICH [6,7], standard deviation can be calculated in one of the following ways:

- At least nine independent determinations in the whole measuring range (e.g., three independent determinations for three concentration levels)
- Six independent determinations of an analyte in standard samples for the concentration level corresponding to the concentration of a real sample
- Six independent determinations of analytes occurring in three different matrices and for two or three concentration levels

According to EURACHEM recommendations [28], one should perform 10 independent determinations and calculate the standard deviation based on these.

The determined method's repeatability can refer both to (1) a very specific analytical method in which matrix composition is specific and defined (e.g., the method of determining analyte *X* concentration in matrix *Y*) and (2) determination methods for a given analyte without specifying matrix composition. In the former case, the standard deviation is calculated based on measurements performed for samples characterized by the same matrix composition. In the latter case, one needs to calculate the standard deviation using the measurements conducted for samples differing in matrix composition.

Intermediate precision is a notion with a wider scope than repeatability because its value is influenced by additional parameters such as [2,3]:

- Personal factors: Different analysts conducting determinations and instability in the work of a given analyst over a specific period
- Instrumental factors: Due to the fact that measurements can be carried out using
  - Different measuring instruments from a given laboratory
  - Standard solutions and reagents coming from different producers, or from different batches
  - Different accessories, for example, different GC columns with the same characteristics but from different producers or different batches

If determining precision uses samples in which analyte concentration is stable, the standard deviation is a sufficient parameter which one may determine precision with. However, in the analysis of samples characterized by different levels of analyte concentration, one should use the relative standard deviation or coefficient of variation. Each of these of two quantities is used to compare repeatability, intermediate precision, or reproducibility.

8.2.6.1   Manners of Estimating the Standard Deviation

Determining intermediate precision, repeatability, and reproducibility is based on calculating the standard deviation for the series of obtained measurement results [33–36]. The simplest means of estimating this parameter is by calculating the relative standard deviation or coefficient of variation and comparing (assessment) the obtained values. Frequently, one can find the statement that if a relative standard deviation (RSD) is smaller than a certain determined limit, then using a given method can yield precise results.

An estimation of standard deviation can be performed using suitable statistical tests:

- With a set point of this parameter—chi square χ^2 test (Section 1.8.4)
- With the value obtained from a statistical assessment of the set of results obtained using a reference method—Snedecor's F test (Section 1.8.5)

Sometimes it is necessary to compare the standard deviation for sets of measurement results obtained using more than two methods. If the number of measurements on which the calculation of standard deviations is based is similar for all methods (equinumerous series of measuring), then one can apply Hartley's F_{max} test (Section 1.8.6).

When the number of results obtained using the compared methods are different, one should compare the calculated standard deviations using Bartlett's test (Section 1.8.7).

If the standard deviations are to be compared for two sets of correlated results, one should use Morgan's test (Section 1.8.8).

Example 8.13

Problem: For the given measurement result series, check (at the level of significance $\alpha = 0.05$) if the calculated standard deviation differs statistically significantly from the set value of the standard deviation.

Apply the χ^2 test.

Data: Results:

11.0	12.0	12.9	12.0	12.5	12.1	14.2	12.1	17.1	12.1	12.4	15.1	12.3	12.0	10.2

$$SD_o = 1.23$$

Solution:

Number of results—n	15
Standard deviation—SD	1.68
χ^2	27.88
χ^2_{crit} $(f = 14, \alpha = 0.05)$	23.68

According to the equation from Section 1.8.4.
Conclusion: Because $\chi^2 > \chi^2_{crit}$, there is a statistically significant difference in variance value.
Excel file: exampl_valid13.xls

Example 8.14

Problem: For the given series of measurement results, check (at the level of significance $\alpha = 0.05$) if the standard deviation values for both the series are statistically significantly different.
Apply Snedecor's F test.
Data: Result series:

	Series 1	Series 2
1	10	11
2	12	11
3	13	13
4	14	11
5	18	13
6	15	12
7	17	

Solution:

	Series 1	Series 2
Number of results—n	7	6
Standard deviation—SD	2.795	0.983
F	7.85	
F_{crit} ($f_1 = 6, f_2 = 5, \alpha = 0.05$)	4.95	

According to the equation from Section 1.8.5.
Solution (2):

F-Test Two Sample for Variances

	Variable 1	Variable 2
Mean	14.143	11.833
Variance	7.810	0.967
Observations	7	6
df	6	5
F	8.079	
$P (F <= f)$ one-tail	0.0183	
F Critica one-tail	4.950	

Conclusion: Because $F > F_{crit}$, there is a statistically significant difference in variance values for the compared series; the series differ in precision.
Excel file: exampl_valid14.xls

Example 8.15

Problem: For the given series of measurement results, check (at the level of signifi-cance $\alpha = 0.05$) if the values of the standard deviation for the given series of results are statistically significantly different.
  Equinumerous series—apply Hartley's F_{max} test.
Data: Result series:

	Series 1	Series 2	Series 3	Series 4	Series 5
1	11	13	10	10	17
2	12	12	13	12	11
3	13	12	14	16	13
4	12	15	12	18	14
5	13	11	13	13	13
6	12	10	14	14	12
7	14	13	11	14	13
8	12	11	12	12	11
9	15	12	17	17	13
10	12	14	14	14	14
11	12	15	17	10	15
12	15	12	12	12	11
13	12	14	11	11	11
14	12	12	12	13	12
15	10	11	14	15	12

Solution:

	Series 1	Series 2	Series 3	Series 4	Series 5
Number of results—n	15	15	15	15	15
Standard deviation—SD	1.36	1.51	2.02	2.38	1.70
F_{max}			3.09		
F_{maxo} $(k = 5, f = 14, \alpha = 0.05)$			4.76		

  According to the equation from Section 1.8.6.
Conclusion: Because $F_{max} < F_{maxo}$, there is no statistically significant difference in variance values for the compared series.
Excel file: exampl_valid15.xls

Example 8.16

Problem: For the given series of measurement results, check (at the level of significance $\alpha = 0.05$) if the values of the standard deviation for a given series of results
are statistically significantly different.
    Not equinumerous—apply Bartlett's test.
Data: Result series:

	Series 1	Series 2	Series 3	Series 4	Series 5	Series 6
1	11	13	10	10	17	10
2	12	12	13	12	11	13
3	13	12	14	16	13	14
4	12	15	12	18	14	12
5	13	11	13	13	13	13
6	12	10	14	14	12	14
7	14	13	11	14	13	11
8	12	11	12	12	11	12
9	15	12	17	17	13	17
10	12	14	14	14		14
11	12	15	17	10		17
12	15		12	12		12
13	12		11	11		
14	12		12			
15	10					

Solution:

	Series 1	Series 2	Series 3	Series 4	Series 5	Series 6
number of results—n	15	11	14	13	9	12
standard deviation—SD	1.36	1.63	2.08	2.53	1.80	2.14
$1/(n-1)$	0.071	0.100	0.077	0.083	0.125	0.091
$(n-1)\cdot\log(SD^2)$	3.701	4.270	8.245	9.672	4.095	7.257
$(n-1)\cdot SD^2$	25.733	26.727	56.000	76.769	26.000	50.250
c			1.04			
SD_o^2			3.845			
Q			51.22			
χ_{crit}^2 $(f=k-1=5, \alpha=0.05)$			11.07			

    According to equations from Section 1.8.7.
Conclusion: Because $Q > \chi_{crit}^2$, there is a statistically significant difference in vari-
ance values for the compared series.
Excel file: exampl_valid16.xls

Example 8.17

Problem: For the given series of measurement results—dependent variables, check (at the level of significance $\alpha = 0.05$) if the values of the standard deviation for the given series of results are statistically significantly different.

Apply Morgan's test.
Data: Result series:

	Series 1	Series 2
1	8.8	9.1
2	9.7	9.8
3	8.9	9.2
4	9.3	9.6
5	8.1	8.2
6	8.9	9.1
7	9.4	9.6
8	9.1	10.1
9	9.2	10.3
10	9.1	9.9
11	8.9	9.7
12	8.2	8.7
13	9.1	9.6

Solution:

r	0.809
SD_1	0.44
SD_2	0.58
L	0.816
t	1.576
t_{crit}	2.201

According to equations from Section 1.8.8.
Conclusion: Because $t < t_{crit}$, there is no statistically significant difference in variance values for the compared series.
Excel file: exampl_valid17.xls

Example 8.18

Problem: In order to determine the values of repeatability, six independent series of measurements were performed for six standard solution samples. In each series five repetitions were made.

Using the obtained measurement results, calculate repeatability for the analytical method.
Data: Result series:

	Series 1	Series 2	Series 3	Series 4	Series 5	Series 6
1	2.54	5.12	7.14	10.2	14.2	17.3
2	2.67	5.16	7.15	10.9	14.8	17.8
3	2.43	5.24	7.34	11.3	13.9	17.2
4	2.65	5.34	7.09	10.2	14.3	17.0
5	2.34	5.02	7.34	10.1	14.4	17.5

Solution: Because the levels of analyte concentrations in the investigated standard solutions samples are different, the calculations should use the values of CV and not SD.

The first step is to check the homogeneity of variances for individual series of results. Because series are equinumerous, one should apply Hartley's F_{max} test (Section 1.8.6).

If variances are homogeneous, repeatability should be calculated as a mean value CV for the given series.

If variances are not homogeneous, one should reject the deviating value (series) and perform the calculations again.

	Series 1	Series 2	Series 3	Series 4	Series 5	Series 6
Number of results—n	5	5	5	5	5	5
Standard deviation—SD	0.142	0.121	0.119	0.532	0.327	0.305
Coefficient of variation—CV,%	5.60	2.34	1.65	5.05	2.28	1.76
F_{max}			11.52			
F_{maxo} $(k = 6, f = 4, \alpha = 0.05)$			29.50			

Conclusion: Because $F_{max} < F_{maxo}$, there is no statistically significant difference in variance values for the compared series. It is possible to calculate repeatability as a mean value CV for the given series.

CV repeatability	3.1%

Other possibilities are to calculate the CV of repeatability according to the following:

$$CV_{repeatability} = \sqrt{\frac{1}{k}\sum_{i=1}^{k} CV_i^2}$$

CV repeatability	3.5%

In this case the checking of homogeneity of variance is not necessary.
Excel file: exampl_valid18.xls

Example 8.19

Problem: To determine the values of repeatability and the intermediate precision of the analytical method, six independent series of measurements for the samples were performed for one standard solution. In each series six repetitions were performed.

Using the obtained measurement results, calculate the values of repeatability and intermediate precision for the analytical method.

Data: Result series, mg/L:

	Series 1	Series 2	Series 3	Series 4	Series 5	Series 6
1	101	103	111	100	103	103
2	104	106	107	102	102	108
3	103	102	104	101	106	102
4	101	105	102	117	103	107
5	100	109	110	115	107	105
6	102	104	105	103	104	103

Solution: The first step is to check the homogeneity of the variances for individual series of results. Because series are equinumerous, one should apply Hartley's F_{max} test.

If variances are homogeneous, repeatability should be calculated as a mean value CV for the given series.

If variances are not homogeneous, one should reject the deviating value (series) and perform the calculations again.

	Series 1	Series 2	Series 3	Series 4	Series 5	Series 6
Number of results—n	6	6	6	6	6	6
Standard deviation—SD, mg/L	1.47	2.48	3.51	7.58	1.94	2.42
F_{max}				26.52		
F_{maxo} ($k = 6, f = 5, \alpha = 0.05$)				18.70		

Conclusion: Because $F_{max} > F_{maxo}$, there is a statistically significant difference in variance values for the compared series. Results from Series 4 should be rejected due to the lack of homogeneity of variances and calculations should be performed again.

Excel file: exampl_valid19a.xls

Data: Result series:

	Series 1	Series 2	Series 3	Series 4	Series 5	Series 6
1	101	103	111	–	103	103
2	104	106	107	–	102	108

3	103	102	104	–	106	102
4	101	105	102	–	103	107
5	100	109	110	–	107	105
6	102	104	105	–	104	103

	Series 1	Series 2	Series 3	Series 4	Series 5	Series 6
Number of results—n	6	6	6	0	6	6
Standard deviation—SD, mg/L	1.47	2.48	3.51	–	1.94	2.42
F_{max}			5.68			
F_{maxo} ($k = 5, f = 5, \alpha = 0.05$)			16.30			

Conclusion: Because $F_{max} < F_{maxo}$, there is no statistically significant difference in variance values for the compared series.

Repeatability was calculated as a mean of SD values for an individual series. Intermediate precision is SD, calculated using all 30 results.

SD repeatability	2.37	mg/L
SD intermediate precision	2.75	mg/L

Excel file: exampl_valid19b.xls

8.2.7 ACCURACY AND TRUENESS

Accuracy is defined as closeness of agreement between a measured quantity value and a true quantity value of a measurand [26].

Trueness is the closeness of agreement between the average of an infinite number of replicate measured quantity values and a reference quantity value [26].

Analysis of these definitions shows that the hitherto existing notion of "accuracy" was replaced by the term "trueness," and the previously applied notion of "accuracy of a single measurement" is now simply "accuracy."

It is trueness that describes the conformity of results obtained using a given analytical method to real (expected) results. It is influenced mostly by the bias of the analytical method.

Accuracy is a combination of trueness and precision. The truer and more precise the results obtained using a given method, the more accurate the result of a single measurement is. Relationships between trueness, precision, and accuracy are presented schematically in Figure 8.3 [4,9].

Of course, other parameters such as linearity and sensitivity also influence the accuracy of an analytical method.

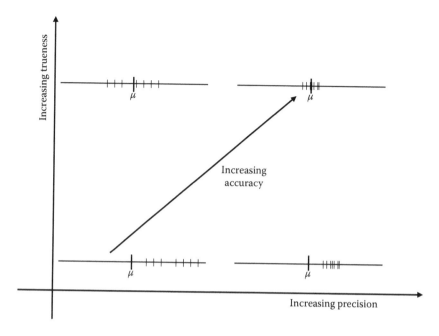

FIGURE 8.3 Relationships between trueness, precision and accuracy. (From Konieczka, P., *Crit. Rev. Anal. Chem.*, 37, 173–190, 2007; Konieczka, P., and Namieśnik, J., eds., *Kontrola i zapewnienie jakości wyników pomiarów analitycznych*, WNT, Warsaw, 2007.)

Trueness and accuracy can be determined using different approaches [33,37–39]:

- Sample analysis of suitable certified reference materials
- Comparison of the obtained result with a result obtained using a reference (primary, definitive) method [40–42]
- Standard addition method

8.2.7.1 Measurement Errors

The notion of accuracy is closely connected with the notion of errors [43]. Depending on the type of errors, their influence on measurements varies.

The value of a single measurement result may differ (and actually always differs) from the expected (real) value. The difference is due to the occurrence of different errors [44]. There are three basic types of errors:

- Gross errors
- Biases
- Random errors

The influence of individual types of errors on a measurement result is presented schematically in Figure 8.4 [9].

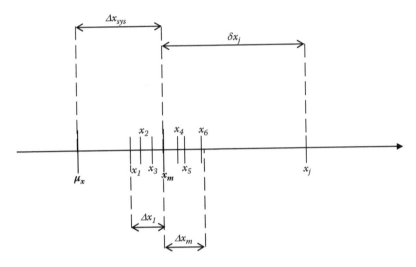

FIGURE 8.4 Influence of individual types of errors on a measurement result. (From Konieczka, P., and Namieśnik, J., eds., *Kontrola i zapewnienie jakości wyników pomiarów analitycznych*, WNT, Warsaw, 2007.)

With regard to the manner of presenting a determination result, one can distinguish:

- Absolute error d_x, which can be described by the dependence

$$d_{x_i} = x_i - \mu_x \tag{8.12}$$

Relative error ε_x, described by the equation

$$\varepsilon_{x_i} = \frac{d_{x_i}}{\mu_x} \tag{8.13}$$

With regard to the source of errors, one can distinguish:

- Methodological errors
- Instrumental errors
- Human errors

The total error of a single measurement result may be divided into three components, as described by the following equation [45]:

$$d_{x_i} = x_i - \mu_x = \Delta x_{sys} + \Delta x_i + \delta x_i \tag{8.14}$$

where

d_{x_i}: Total error of a measurement result
x_i : Value of a measurement result
μ_x: Expected value
Δx_{sys}: Bias
Δx_i: Random error
δx_i : Gross error

For a measurement series (at least three parallel analyte determinations in the same sample), there is a high probability of detecting a result(s) with a gross error.

Gross error is characterized by the following properties:

- It is the result of a single influence of a cause acting temporarily.
- It appears only in some measurements.
- It is a random variable—however, one with unknown distribution and unknown expected value.
- It is the easiest to detect, and therefore to eliminate.
- It assumes both positive and negative values (unlike bias).
- The cause of its occurrence can be, for example, a mistake in instrument reading or a mistake in calculations.

There are many known ways of detecting results with gross errors. Each of them is applied in certain specific conditions.

Methods of gross error determination are described in Chapter 1.

Figure 8.5 schematically presents the selection criteria for a suitable manner of action in detecting and rejecting results with gross errors, often described as "outliers" [9].

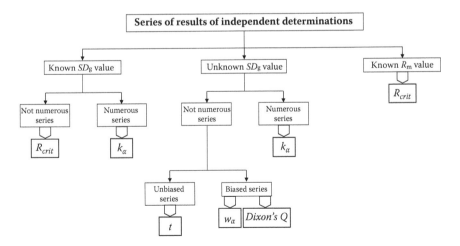

FIGURE 8.5 Selection criteria for a suitable manner of action in detecting and rejecting results with gross errors, often described as "outliers." (From Konieczka, P., and Namieśnik, J., eds., *Kontrola i zapewnienie jakości wyników pomiarów analitycznych*, WNT, Warsaw, 2007.)

After eliminating results with gross errors, the trueness of the obtained final determination (most often the mean value of the measurement series) is influenced by biases or random errors.

The determination of biases is one way to determine the trueness of an analytical method.

Table 8.5 presents specific methods of bias determination [9].

TABLE 8.5

Basic Information Concerning Methods of Bias Determination

Bias Type	Requirements	Course of Action
Constant	Samples of two standards (reference materials) with different analyte content	Series determinations for two standard samples (reference material samples) with different analyte content, using the developed method. Constant bias a_{sys} is determined according to the following formula:

$$a_{sys} = \frac{\mu_{1x} x_{1m} - \mu_{2x} x_{2m}}{\mu_{1x} - \mu_{2x}} \qquad (8.15)$$

where

μ_{1x}, μ_{2x} are the expected values for two standard samples
x_{1m}, x_{2m} are the mean values determined for standard samples

Variable	Sample and sample with standard addition	Series determination with the use of the developed method for sample and sample with standard addition. The correction multiplier value is determined according to the formula:

$$B = \frac{C_{st}}{x_{mCst} - x_m} \qquad (8.16)$$

where

C_{st}: Expected value increase of analyte concentration due to standard addition

x_m, x_{mCst}: Mean values determined for sample and sample with standard addition

The value of variable bias is determined according to the following equation:

$$b_{sys} = \frac{1 - B}{B} \qquad (8.17)$$

Variable	Samples of two standards (reference materials) Reference method	Two series of determination for two standard samples with the use of the reference method and the developed method. The correction multiplier value is determined according to the following formula:

$$B = \frac{x_{2m(ref)} - x_{1m(ref)}}{x_{2m} - x_{1m}} \qquad (8.18)$$

(Continued)

TABLE 8.5 (CONTINUED)
Basic Information Concerning Methods of Bias Determination

Bias Type	Requirements	Course of Action
		where
		$x_{1m(ref)1}$, $x_{2m(ref)}$: Mean values determined for the 1st and 2nd standard using the reference method
		x_{1m}, x_{2m}: Mean values determined for the 1st and 2nd standard using the developed method
		The value of variable bias is determined according to the Equation 8.17
Constant and variable	Series samples with different analyte content, Reference method	Series determination for samples with different analyte content with the use of the reference method and the developed method
		The relationship between results obtained by the reference method (OY-axis) and results obtained by the developed method (OX-axis) is determined
		Regression parameters of the regression line $Y = b \cdot X + a$ are determined according to the Equations 1.63 and 1.64
		The values of constant bias and variable bias are determined according to the following formulas:
		(a) Constant bias:

$$a = -a_{sys} b \tag{8.19}$$

therefore,

$$a_{sys} = -\frac{a}{b} \tag{8.20}$$

(b) Variable bias:

$$b = B \tag{8.21}$$

therefore,

$$b_{sys} = \frac{1}{b} - 1 \tag{8.22}$$

Source: Konieczka, P., and Namieśnik, J., eds., *Kontrola i zapewnienie jakości wyników pomiarów analitycznych*, WNT, Warsaw, 2007 (in Polish).

A determination result (arithmetical mean of a series of parallel measurements) can only have a bias and random error according to the following dependence [45]:

$$d_{x_m} = x_m - \mu_x = \Delta x_{sys} + \Delta x_m \qquad (8.23)$$

where

d_{x_m}: Total error of a determination result (arithmetical mean of the series of measurements)

x_m: Mean value of the series of measurement results

μ_x: Expected value

Δx_{sys}: Bias

Δx_m: Random error

If the determined bias refers to an analytical method, then with a large number of conducted measurements, the random error is negligibly small with relation to the bias, when $n \to \infty$, than $s \to 0$.

In this case, the following dependence is true [45]:

$$d_{x_{met}} = E(x_{met}) - \mu_x = \Delta x_{sys} \qquad (8.24)$$

where

$d_{x_{met}}$: Total error of a determination result for the applied analytical method

$E(x_{met})$: Value of a determination obtained as a result of a given analytical method used (expected value for a given analytical method)

μ_x: Expected value (real)

Δx_{sys}: Bias

In this way, the bias of an analytical method is determined. The occurrence of bias makes a given series of measurement (analytical method) results differ from the expected value by a constant value—hence they are either overstated or understated.

One may differentiate between two types of bias:

- A constant bias, whose value is not relative to analyte concentration levels–a_{sys}
- A variable bias, whose value depends (most often linearly) on analyte concentration levels—$b_{sys}\mu_x$

Bias is described by the following dependence:

$$\Delta x_{sys} = a_{sys} + b_{sys}\mu_x \qquad (8.25)$$

Assuming that the value of a random error is negligibly small compared to the bias value, one can present the following dependence:

$$x_m = \mu_x + \Delta x_{sys} = \mu_x + a_{sys} + b_{sys}\mu_x = a_{sys} + (1 + b_{sys})\mu_x \qquad (8.26)$$

Only after rejecting results with a gross error and determining biases (regarding their values and correcting the determination result) can a result have a random error. Its value influences the precision of the obtained results.

The trueness or accuracy can be determined using different techniques [37–39].

One of them is comparing the obtained measurement value with the value obtained resulting from a reference method for which the obtained results are treated as accurate. In this case, one can compare both results visually, but it is more metrologically correct to use Student's t test (Section 1.8.9) for the significance of differences between two results. Of course, this test can only be applied when the compared methods do not differ in a statistically significant manner with respect to precision (Snedecor's F test—Section 1.8.5).

However, when the result of the Snedecor's F test application is negative (standard deviations for the series of measurements obtained by the compared analytical methods differ statistically and significantly), one may use for "poor" (small) result series the "approximate test" of Cochran's C and the Cox tests (Section 1.8.10) or the Aspin and Welch tests (Section 1.8.11).

Another manner (most often applied) to determine the trueness or accuracy is the analysis of a reference material sample (or still better, samples of the certified reference material) using the investigated analytical method.

According to the general definition, reference material is characterized by a constant and strictly defined analyte concentration and with a known concentration determination uncertainty [26]. Of course, it is not always possible to use reference material samples for precisely satisfying given needs. In case of its inaccessibility, one should prepare a standard solution by adding a strictly specific quantity of analyte into the investigated sample and subject it to determination. In each case, however, one should perform independent determinations for a blank sample and correct the result for the sample with the known analyte concentration by the obtained measurement result.

In order to test whether the obtained measurement value differs in a statistically significant manner from the certified value (expected value), one should apply Student's t test (Section 1.8.9).

An insignificant difference between two obtained results may also be tested by using the method of calculating the ratio between the obtained results and uncertainties of their determination.

One determines the ratio of the obtained means (if the values did not differ between themselves, the ratio should be 1) and values of uncertainty for such a determined quantity. The inference is as follows: If the interval of a determined ratio ± the uncertainty of its determination ($R \pm U$) includes 1, one should infer that the compared mean values do not differ in a statistically significant manner.

Using obtained values, one should calculate the value of the R ratio according to the following formula

$$R = \frac{x_{det}}{x_{ref}} \tag{8.27}$$

and then the uncertainty U, using a dependence described by the following equation

$$U = k \frac{\sqrt{u_{det}^2 + u_{ref}^2}}{\left(\dfrac{x_{det} + x_{ref}}{2}\right)} \tag{8.28}$$

where

U: Expanded uncertainty for determined relation

k: Coverage factor whose value depends on the accepted level of probability (most often 95 percent for which $k = 2$)

There is also another manner based on the comparison of values calculated from the dependence which can be presented using the following expressions:

$$\left| x_{det} - x_{ref} \right| \tag{8.29}$$

$$2\sqrt{u_{(x_{det})}^2 + u_{(x_{ref})}^2} \tag{8.30}$$

where

x_{det}: Value of a determination result

x_{ref}: Reference value

$u_{(x_{det})}$: Uncertainty of a determination result

$u_{(x_{ref})}$: Uncertainty of a reference value

Inference in this instance is as follows:

- If the inequality

$$\left| x_{det} - x_{ref} \right| < 2\sqrt{u_{(x_{det})}^2 + u_{(x_{ref})}^2}$$

  occurs, then the result is deemed to be in conformity with the reference value.
- However, when the following dependence is true:

$$\left| x_{det} - x_{ref} \right| \geq 2\sqrt{u_{(x_{det})}^2 + u_{(x_{ref})}^2}$$

  then the result is acknowledged to not be in conformity with the reference value.

This manner of inference is based on comparing differences between two results with the expanded uncertainty (for $k = 2$) calculated using the uncertainty for the compared values.

According to recommendations by the ICH [6,7], determining trueness should be carried out using at least nine parallel determinations at three different analyte concentration levels (at least three determinations per each level of concentration). The calculated trueness should be presented as the percentage of recovery of the expected value or as a difference between the mean and the expected value together with the given confidence interval.

EURACHEM [28] recommends 10 parallel determinations for a blank sample and the same number of determinations for reference material samples. The mean obtained for blank samples is then deduced from the mean obtained for the reference material, and so the corrected value is compared against the certified value.

It is also recommended to perform a series of measurements for the reference material using a so-called primary method, characterized by a null value of bias. In this case, the corrected mean obtained for the investigated method is compared with the one obtained by the primary method.

Example 8.20

Problem: In the given series of measurement results, check if there is a result with a gross error. Apply the confidence interval method, after the initial outlier rejection. Assume the value $\alpha = 0.05$.
Data: Result series, mg/dm^3:

	Data
1	8.8
2	7.8
3	9.2
4	9.5
5	6.3
6	8.2
7	9.1
8	8.8

α	0.05

Solution:

x_{min}	6.3
x_{min+1}	7.8
x_{max}	9.5
x_{max-1}	9.2
t_{crit}	2.447

Initially the result x_{min} was rejected.

x_m	8.77	mg/dm³
SD	0.59	mg/dm³

$$g = x_m \pm t_{crit}\sqrt{\frac{n}{n-2}}SD$$

g	8.77 ± 1.67 mg/dm³
	(7.10 ÷ 10.44) mg/dm³

Conclusion: The value x_{min} lies outside the determined confidence interval—hence, it has a gross error.
Excel file: exampl_valid20.xls

Example 8.21

Problem: Using the data from the Example 8.20, apply the confidence interval method, without the initial outlier rejection. Assume the value α = 0.05.
Data: Result series, mg/dm³:

	Data
1	8.8
2	7.8
3	9.2
4	9.5
5	6.3
6	8.2
7	9.1
8	8.8

α	0.05

Solution:

x_{min}	6.3
x_{max}	9.5
w_α	1.87

x_m	8.46	mg/dm³
SD	1.0	mg/dm³

$$g = x_m \pm w_\alpha \cdot SD$$

g	8.46 ± 1.93 mg/dm³
	(6.53 ÷ 10.39) mg/dm³

Conclusion: The value x_{min} lies outside the determined confidence interval—hence, it has a gross error. It should be rejected and the values of x_m and SD should be calculated for the new series of data.

x_m	8.77	mg/dm³
SD	0.59	mg/dm³

Excel file: exampl_valid21.xls

Example 8.22

Problem: Using the data from the Example 8.20, apply Dixon's Q test. Assume the value $\alpha = 0.05$.
Data: Result series, mg/dm³:

	Data
1	8.8
2	7.8
3	9.2
4	9.5
5	6.3
6	8.2
7	9.1
8	8.8

α	0.05

Solution:

Number of results	8
Range—R	3.20
Q_1	0.469
Q_n	0.094
Q_{crit}	0.468

According to equations from Section 1.8.3.

Conclusion: Because $Q_1 > Q_{crit}$, the value x_{min} has a gross error. It should be rejected and one the values of x_m and SD should be calculated for the new series of data.

x_m	8.77	mg/dm³
SD	0.59	mg/dm³

Excel file: exampl_valid22.xls

Example 8.23

Problem: In a given series of measurement results, check if there are any results with a gross error. Apply the confidence interval method. Assume the value $\alpha = 0.05$.

Data: Result series, ppm:

	Data
1	13.2
2	13.7
3	13.9
4	14.1
5	13.4
6	13.2
7	13.4
8	13.7
9	14.2
10	11.3
11	13.4
12	13.2
13	13.8
14	14.2
15	14.2
16	15.8
17	15.4

18	13.2
19	13.3
20	13.7
21	13.7
22	13.8
23	13.2
24	14.1
25	14.2
26	13.9
27	13.2
28	13.6
29	13.4
30	13.7
31	14.1
32	14.0
33	13.8

α	0.05

Solution:

x_m	13.73	ppm
SD	0.72	ppm

k_α	1.65

$$g = x_m \pm k_\alpha \cdot SD$$

g	13.73 ± 1.19 ppm
	(12.53 ÷ 14.92) ppm

	Data
1	13.2
2	13.7
3	13.9
4	14.1
5	13.4
6	13.2
7	13.4

8	13.7
9	14.2
10	11.3 outlier
11	13.4
12	13.2
13	13.8
14	14.2
15	14.2
16	15.8 outlier
17	15.4 outlier
18	13.2
19	13.3
20	13.7
21	13.7
22	13.8
23	13.2
24	14.1
25	14.2
26	13.9
27	13.2
28	13.6
29	13.4
30	13.7
31	14.1
32	14.0
33	13.8

Conclusion: Results 10, 16, and 17 lie outside the determined confidence interval; hence, they have a gross error.

After their rejection the values of x_m and *SD* were calculated again.

x_m	13.68	ppm
SD	0.36	ppm

Excel file: exampl_valid23.xls

Example 8.24

Problem: Check if there are results with a gross error in a given series of measurement results. Results of measurements were obtained using a method for which the standard deviation method had been determined.

Apply the critical range method. Assume the value $\alpha = 0.05$.

Data: Result series, ppb:

	Data
1	113
2	125
3	120
4	127
5	115
6	118
7	117
8	134
9	124

α	0.05
SD_g	4.5

Solution:

x_{min}	113
x_{min+1}	115
x_{max}	134
x_{max-1}	127
z (Appendix Table A.3)	4.39

R	21.0	ppb
R_{crit}	19.8	ppb

$$R_{crit} = z \cdot SD_g$$

Conclusion: Because $R > R_{crit}$, a result x_{max} is considered to be an outlier, and new calculations for the new series should be done.

Data (2): Result series, ppb:

	Data
1	113
2	125
3	120
4	127
5	115

6	118
7	117
8	–
9	124

Solution (2):

x_{min}	113
x_{min+1}	115
x_{max}	127
x_{max-1}	125
z (*Appendix Table A.3*)	4.29

| R | 14.0 | ppb |
| R_{crit} | 19.3 | ppb |

Conclusion: Because $R < R_{crit}$ there are no more outliers in the series, so the values of x_m and SD could be calculated.

| x_m | 121 | ppb |
| SD | 6.65 | ppb |

Excel file: exampl_valid24.xls

Example 8.25

Problem: Check if there is a result with a gross error in a given series of measurement results. The results were obtained using a method for which a standard deviation had been determined beforehand.

Apply the confidence interval method. Assume the value $\alpha = 0.05$.
Data: Result series, ng/g:

	Data
1	55.2
2	54.8
3	56.1
4	56.7
5	53.1
6	57.1
7	54.2

8	55.5
9	57.0
10	56.8
11	53.3
12	51.9
13	52.1
14	51.7
15	54.2
16	54.3
17	55.5

α	0.05
SD_g	1.9

Solution:

x_{min}	51.7
x_{min+1}	51.9
x_{max}	57.1
x_{max-1}	57.0
k_{α}	1.65

Result x_{min} was initially rejected.
The confidence interval value was calculated for the new series.

x_m	54.9	ng/g

$$g = x_m \pm k_\alpha \cdot SD_g \sqrt{\frac{n}{n-1}}$$

g	54.9 ± 3.2 ng/g
	(51.6 ÷ 58.1) ng/g

Conclusion: An initially rejected result x_{min} lies in the determined confidence interval. It has been included in the series and the values of x_m, i, and SD were calculated again.

x_m	54.7	ng/g
SD	1.8	ng/g

Excel file: exampl_valid25.xls

Example 8.26

Problem: Determinations were made for 25 samples, performing three parallel determinations per each sample. Using the data-obtained measurement results, check them for the occurrence of outliers.

Apply the critical range method. Assume the value $\alpha = 0.05$.

Data: Result series, ppm:

Sample	Result 1	Result 2	Result 3
1	3.01	3.33	3.35
2	3.11	3.04	3.13
3	3.65	3.45	3.41
4	3.23	3.45	3.12
5	3.22	3.13	3.33
6	3.28	3.41	3.62
7	3.45	3.12	3.04
8	3.65	3.07	3.45
9	3.01	3.08	3.99
10	3.14	3.52	3.88
11	3.11	3.71	3.12
12	3.65	3.74	3.07
13	3.23	3.32	3.04
14	3.67	3.22	3.2
15	3.98	3.11	3.44
16	3.56	3.41	3.49
17	3.33	3.49	3.82
18	3.11	3.51	3.72
19	3.23	3.82	3.23
20	3.41	3.01	3.67
21	3.21	3.01	3.98
22	3.48	3.37	3.56
23	3.6	3.62	3.33
24	3.62	3.08	3.62

α	0.05
z_α	1.96

Solution:

Sample	R_i	Conclusion
1	0.34	OK
2	0.09	OK
3	0.24	OK
4	0.33	OK
5	0.20	OK
6	0.34	OK
7	0.41	OK
8	0.58	OK
9	0.98	Outlier
10	0.74	OK
11	0.60	OK
12	0.67	OK
13	0.28	OK
14	0.47	OK
15	0.87	OK
16	0.15	OK
17	0.49	OK
18	0.61	OK
19	0.59	OK
20	0.66	OK
21	0.97	Outlier
22	0.19	OK
23	0.29	OK
24	0.54	OK

R_m	0.49
R_{crit}	0.96

$$R_i = x_{max_i} - x_{min_i}$$

$$R_{crit} = z_\alpha \cdot R_m$$

Conclusion: For series 9 and 21, $R_i > R_{crit}$ results should be rejected as outliers.
Excel file: exampl_valid26.xls

Example 8.27

Problem: Analyte concentrations were determined in two standard solution samples, with seven parallel determinations performed per sample. A second standard solution was obtained by double dilution of the first standard solution.

Using the obtained result series, determine the value of the constant bias a_{sys}.

Data: Result series, ppm:

	Results	
	Series 1	Series 2
1	10.01	5.33
2	10.11	5.04
3	10.07	5.11
4	10.23	5.45
5	10.22	5.13
6	10.28	5.41
7	10.23	5.12

x_{1st}	10.0
x_{2st}	5.0

Solution:

x_{1m}	10.16
x_{2m}	5.23
k	2

$$k = \frac{x_{1st}}{x_{2st}}$$

$$a_{sys} = \frac{kx_{1m} - x_{2m}}{k - 1}$$

a_{sys}	0.290	ppm

Excel file: exampl_valid27.xls

Example 8.28

Problem: Analyte concentrations were determined in a real sample and in a real sample with the standard addition. For each of the samples, six parallel measurements were made.

Using the data-obtained result series, determine the value of the variable bias b_{sys}. Using the calculated value of the correction multiplier, correct the values obtained for the real sample.

Data: Result series, ppm:

	Results	
	Series 1	Series 2
1	33.4	57.2
2	33.8	56.9
3	34.2	58.2
4	33.9	57.5
5	33.1	58.8
6	33.9	58.5

C_{st}	25.0

Solution:

x_m	33.72
x_{mCst}	57.85

$$B = \frac{C_{st}}{x_{mC_{st}} - x_m}$$

$$b_{sys} = \frac{1-B}{B}$$

$$x_{m(corr)} = B \cdot x$$

B	1.036
b_{sys}	−0.035
$x_{m(corr)}$	34.93

Excel file: exampl_valid28.xls

Example 8.29

Problem: Analyte concentrations were determined in two real samples, using an investigated method and the reference method.

For each of the samples, eight parallel measurements were made, using both methods.

Using the obtained result series, determine the value of the variable bias b_{sys}. Using the calculated value of the correction multiplier, correct the values obtained using the validated method.

Data: Result series, ppb:

	Reference Method		Validated Method	
	Sample 1	Sample 2	Sample 1	Sample 2
	x_{1ref}	x_{2ref}	x_1	x_2
1	746	945	765	967
2	740	947	772	980
3	753	956	758	978
4	758	960	768	984
5	743	948	783	974
6	750	955	749	984
7	746	960	777	975
8	755	966	769	988

Solution:

$x_{1m(ref)}$	748.88
$x_{2m(ref)}$	954.63
x_{1m}	767.63
x_{2m}	978.75

$$B = \frac{x_{2m(ref)} - x_{1m(ref)}}{x_{2m} - x_{1m}}$$

$$b_{sys} = \frac{1-B}{B}$$

$$x_{m(corr)} = B \cdot x$$

B	0.975
b_{sys}	0.026
$x_{1m(corr)}$	748.08
$x_{2m(corr)}$	953.83

Excel file: exampl_valid29.xls

Example 8.30

Problem: Analyte concentrations were determined in fifteen real samples using the validated method and the reference method.

For each of the samples, three parallel measurements were made using each of the methods and the mean values were presented.

Using the obtained data, determine the variable bias b_{sys} and the constant bias a_{sys}. Apply the linear regression method.

Data: Result series, ppb:

	Validated Method	Reference Method
	x	x_{ref}
1	46.9	45.7
2	88.5	86.9
3	101	97.8
4	79.4	77.2
5	21.2	19.6
6	12.3	10.9
7	109	103
8	59.3	56.8
9	57.3	56.2
10	47.2	44.2
11	39.3	35.2
12	38.1	37.2
13	27.3	26.8
14	90.2	89.3
15	111	106

Solution:

$$x_{ref} = b \cdot x + a$$

$$a_{sys} = -\frac{a}{b}$$

$$b_{sys} = \frac{1}{b} - 1$$

a	−0.589
b (B)	0.972
a_{sys}	0.607
b_{sys}	0.0292

Graph:

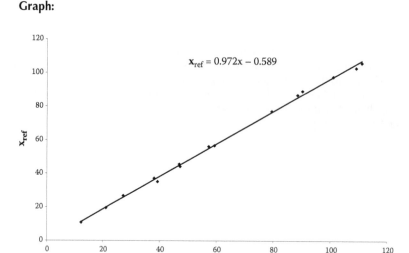

$$x_{ref} = 0.972x - 0.589$$

Excel file: exampl_valid30.xls

8.2.8 ROBUSTNESS AND RUGGEDNESS

The robustness of a method is determined in order to find the influence of slight fluctuations of conditions in a given analytical method on the result of final determination. Robustness influences the manner of conducting measurements using a given analytical method. The greater the influence of slight changes in parameters of the measurement process on final determination results, the greater the attention one should pay to maintaining these parameters at a stable level. It is a parameter concerning changes in internal conditions [46,47].

However, ruggedness (flexibility) is a parameter describing the usefulness of a given analytical method in different conditions, and can be estimated based on reproducibility [46,47].

Similarly to the reproducibility of an analytical method, its robustness and ruggedness are also determined in interlaboratory studies, although the influence of fluctuations from some measurement conditions (in a method subjected to validation) may be conducted in one laboratory (e.g., the influence of fluctuations in temperature, changes in purity and types of reagents, pH fluctuations, conditions of chromatographic isolations) [46,47].

These parameters can be calculated based on a study of changes in the standard deviation of the measurement series using a given analytical method, and slightly fluctuating the parameters of the applied analytical method.

8.2.9 UNCERTAINTY

Uncertainty is not considered a basic validation parameter, but it should be presented in the final method validation report. Based on the estimated uncertainty value, one can determine the usefulness of a given analytical method for a given determination. Determination of a combined uncertainty for an investigated analytical method (most often expressed as a percentage of the determined value) makes it possible to know the quality of results obtained with a given method. The exact characterization of this parameter, together with a description of its determination, is presented in Chapter 5.

Example 8.31

General problem: An analytical procedure was developed, indicating the content of total mercury in samples of muscle tissue of great cormorants (*Phalacroxorac carbo*) with the use of atomic absorption spectroscopy (cold vapor technique). The validation process method was conducted, determining the appropriate validation parameters.

Problem 1: Determine the *selectivity* of the *CV-AAS* method.

Solution: In the case of the cold vapor technique, mercury is released from the analyzed sample, and then (after an eventual reduction to atomic mercury), it is trapped on the gold bed as an amalgam. After this step, the amalgam is heated to 600°C and the released atomic mercury is directed through the air stream to the absorption cell, in which an absorption measurement is conducted, with a wavelength of 253.7 nm, sent by hollow mercury cathode lamp.

Conclusion: Such a measurement method guarantees high selectivity for indicating mercury for two reasons:

1. The amalgamation reaction is a selectivity reaction for mercury.
2. The absorption measurement is realized using a characteristic wavelength for mercury.

Problem 2: Based on measurement results for the series of standard solutions, determine the *linearity* of the method.

Data: Results:

Unit					
Content of Hg, ng	20	40	60	80	100
Signal	33.5	67.3	99.5	142.1	167.6
	34.1	66.4	98.3	137.8	175.2
	35.2	63.8	99.1	140.1	170.2
	32.8	68.1	100.2	136.2	169.3
	33.9	66.6	95.6	138.0	171.1

Solution: Before constructing the calibration curve, the homogeneity of variation for the results of the series being analyzed should be checked. For this, Hartley's F_{max} test was applied with a significance level of $\alpha = 0.05$.

Content of Hg	20	40	60	80	100
No results—n	5	5	5	5	5
Signal, mean	31.58	62.03	92.12	129.03	158.90
Standard deviation—SD	0.880	1.62	1.78	2.29	2.84
CV	2.79%	2.61%	1.93%	1.77%	1.79%
F_{max}			10.40		
F_{maxo}			25.20		

Conclusion: There are no statistically significant differences in variation values. **Excel file:** exampl_valid31_1.xls

Due to no statistically significant differences in variation for the compared series, a calibration curve was constructed and its regression parameters were determined.

n	25
slope—b	1.730
intercept—a	−2.1
residual standard deviation—SD_{xy}	2.7
standard deviation SD_b	0.019
standard deviation SD_a	1.3
regression coefficient—r	0.9986

Graph:

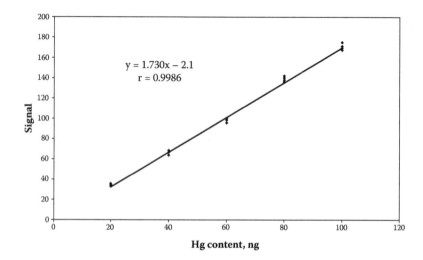

Excel file: exampl_valid31_2.xls
Conclusion: A high value of the regression coefficient r, with the fulfillment of the equal distribution of the standard in the range of the calibration line, requires a high linearity procedure.
Problem 3: Based on the series of measurement results for the standard solutions with the three lowest mercury content levels (20, 40, and 60 ng), determine the *LOD* value, the *LOQ* value, and the range.
    Additionally, check the correctness of the *LOD* determination.
Solution:

n	15
Slope—b	1.616
Intercept—a	1.65
Residual standard deviation—SD_{xy}	1.4
Standard deviation SD_b	0.023
Standard deviation SD_a	0.97
Regression coefficient—r	0.9987

The *LOD* value was determined using the following equation:

$$LOD = \frac{3.3\,s}{b}$$

LOD (SD_{xy})	2.9 ng
LOD (SD_a)	2.0 ng
LOD (mean)	2.5 ng

Graph:

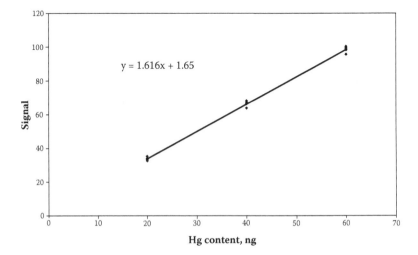

y = 1.616x + 1.65

The correctness of LOD determination was made according to the following equations:

$$10 \cdot LOD > c_{min}$$

$$LOD < c_{min}$$

where $c_{min} = 20$ ng.
Conclusion: The determined LOD value is correct.
Based on the relationship

$$LOQ = 3 \, LOD$$

the LOQ value was calculated to be

$$LOQ = 7.4 \text{ ng}$$

Whereas the range was presented as

$$7.4 \div 100 \text{ ng}$$

Excel file: exampl_valid31_2.xls
Problem 4: Based on the series of results for the three real samples (lyophilized muscle tissue of the great cormorant), calculate the *repeatability*.
Data: Measurement results for individual samples:

	Sample 1		
	Sample Mass, mg	Hg Content, ng	Hg Concentration, ppm
1	30.4	64.14	2.11
2	32.5	71.82	2.21
3	33.8	78.42	2.32
4	30.7	66.62	2.17
5	31.2	70.20	2.25
6	37.3	91.01	2.44
7	35.1	79.68	2.27
	Mean		2.25

	Sample 2		
	Sample Mass, mg	Hg Content, ng	Hg Concentration, ppm
1	25.2	78.37	3.11
2	27.8	85.90	3.09
3	28.3	90.84	3.21
4	22.8	71.82	3.15

5	21.9	72.93	3.33
6	24.9	84.91	3.41
7	25.0	81.50	3.26
	Mean		**3.22**

	Sample 3		
	Sample Mass, mg	**Hg Content, ng**	**Hg Concentration, ppm**
1	21.1	83.77	3.97
2	20.7	80.11	3.87
3	22.3	78.94	3.54
4	24.4	92.48	3.79
5	20.9	81.93	3.92
6	19.7	72.69	3.69
7	20.5	74.00	3.61
	Mean		**3.77**

Solution: Before performing the calculation, in order to indicate precision, one should check whether in the measurement results series there are no outliers. For this, the Dixon's Q test was applied (with a significance level $\alpha = 0.05$).

	Sample 1	Sample 2	Sample 3
No of results—n	7	7	7
Range—R	0.33	0.32	0.43
Q_1	0.182	0.062	0.162
Q_n	0.363	0.250	0.116
Q_{crit}	0.507		

Conclusion: In the series of measurement results, there are no outliers.
Excel file: exampl_valid31_3.xls

Determinations were conducted for three different real samples; therefore, before calculating the repeatability value (as the mean of the coefficient variation for the results of the three series results), the homogeneity of the variation should be checked for the series of results to be analyzed. Hartley's F_{max} test was applied with this aim (a significance level of $\alpha = 0.05$ was chosen).

	Sample 1	Sample 2	Sample 3
No results—n	7	7	7
standard deviation—SD	0.107	0.118	0.162
$CV,\%$	4.76%	3.67%	4.30%
F_{max}		1.68	
F_{maxo}		8.38	

Conclusion: There are no statistically significant differences in variation values. The calculated repeatability value, however, can be calculated as a mean value from the coefficient of variation—CV, counted for three series:

$CV_{repeatability}$	4.24%

Excel file: exampl_valid31_4.xls
Problem 5: Based on the results determined for certified reference material samples (BCR-463—Tuna fish: total Hg and methylmercury), determine the *trueness* value (as a recovery value).
Data: Results are given as (ng/mg):

	Data
1	2.678
2	2.753
3	2.516
4	2.970
5	2.918

	Value	U	k
CRM	2.85	0.16	2

Solution:

Mean	2.77
SD	0.184
U	0.16
%R	97.1%
U (k = 2)	8.2%

where the expanded uncertainty of the recovery value is calculated in accordance with the following equation

$$U = k \frac{\sqrt{\left(u_{CRM}^2 + u_{det}^2\right)}}{\left(\dfrac{X_{CRM} + X_{det}}{2}\right)}$$

Trueness = 97.1 ± 8.2

Graph:

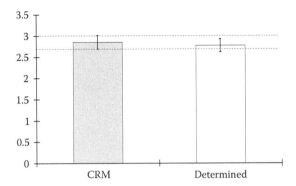

Conclusion: The results obtained with the use of the developed method are correct.

Excel file: exampl_valid31_5.xls

Problem 6: Estimate an uncertainty value for the determination results of the total mercury content in real samples, obtained with the use of the elaborated method.

Solution: As the main components of the uncertainty budget, the following were recognized: The uncertainty value resulting from the calibration curve, the uncertainty value related to the unrepeatability of the measurement results, as well as the uncertainty value from the indication of trueness.

The estimation of the combined uncertainty value was conducted using the following relationship:

$$u_{smpl} = \sqrt{u_{cal}^2 + u_{rep}^2 + u_{true}^2}$$

where

u_{smpl}: Combined relative standard uncertainty for determined results for the real sample

u_{cal}: Relative standard uncertainty related to the calibration step

u_{rep}: Relative standard uncertainty related to repeatability of measurement results

u_{true}: Relative standard uncertainty related to indicating trueness

The determination of the standard uncertainty value related to the calibration step (preparation of the series of standard solutions, conducting measurements for the series of standard solutions, an approximation of measurement points of the calibration line using line regression) was conducted on the basis of the calibration parameters. Calculations were conducted for minimal weighted masses for each of the analyzed real samples.

	Sample 1	Sample 2	Sample 3
No results—n	7	7	7
Minimum Hg content, ng	64.14	71.82	72.69
Hg, concentration, ppm	2.25	3.22	3.77
u_{cal}, %	1.1	0.96	0.96
u_{rep}, %	1.8	1.4	1.6
u_{true}, %		4.2	
u_{smpl}, %	4.7	4.5	4.6
u_{smpl}, ppm	0.11	0.15	0.17
U_{smpl} ($k = 2$), ppm	0.21	0.29	0.35
U_{smpl} ($k = 2$), %	9.4	9.1	9.3

Graph:

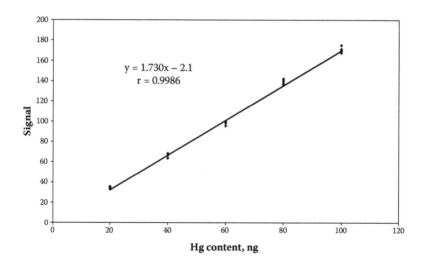

$$y = 1.730x - 2.1$$
$$r = 0.9986$$

Conclusions: The estimated expanded uncertainty value for measurement results for real samples does not exceed 10 percent and allows for the notation of measurement results as follows:

Sample 1: 2.25 ± 0.21 ppm
Sample 2: 3.22 ± 0.29 ppm
Sample 3: 3.77 ± 0.35 ppm

Excel files: exampl_valid31_6.xls
exampl_valid31_7.xls

8.3 CONCLUSION

Validation of an analytical method should be finished with a final report containing [2,9]:

- Subject matter and the purpose of the analytical method (applicability range).
- Metrological principles.
- Type of the applied analyte(s) and matrix composition.
- List of all reagents, standards, and reference materials used, together with precise specification (purity, quality, producer, and, in case of laboratory synthesis, a detailed description of this synthesis).
- Description of the methods used for testing the purity of the substances used and the quality of standards.
- Safety requirements.
- A plan describing the means of transferring the method from laboratory conditions to routine measurements.
- Parameters of the method.
- A list of critical parameters whose slight fluctuations can significantly influence a final determination result—parameters resulting from determination of the analytical method's ruggedness.
- List of all types of laboratory instrumentation together with their characteristic features (dimensions, precision class, etc.), block schemes in case of complicated instrument kits.
- Detailed description of the conditions for conducting the analytical method.
- Description of statistical conduct together with the enclosed suitable equations and calculations.
- Description of the method in order to inspect its quality in routine analyses.
- Suitable figures and graphs, for example, chromatograms and calibration curves.
- Conformity of the determined validation parameters with the assumed limits.
- The uncertainty of a measurement result.
- Criteria that one should fulfill in revalidation.
- Full name of the person which conducted the validation process.
- List of literature used.
- Recapitulation and conclusions.
- Confirmation and signature of the person responsible for the test and confirmation of the validation.

Example 8.32

Problem: Based on the validation parameters indicated for the analytical procedure in Example 8.31, create a validation report.
Solution: Seabirds are useful bioindicators of coastal and marine pollution. Marine birds spend a significant portion of their lives in coastal or marine environments

and are exposed to a wide range of chemicals, because most occupy higher trophic levels, making them susceptible to bioaccumulation of pollutants.

Great cormorants (*Phalacrocorax carbo*) were used as bioindicators for mercury contamination due to their specific feeding habits, wide geographical ranges, and long life span.

The analytical procedure is intended for determining whole mercury content in muscle tissue samples from great cormorants.

Measurements of the content of total mercury will be performed using the cold vapor AAS technique.

A sample is thermally decomposed, mercury is further atomized, and free mercury vapor in the generated gas is collected by a mercury collection agent (gold-coated diatomite particle support) in the form of a gold amalgam. The mercury collection agent is then heated up to 600°C to release atomic mercury. The released mercury is detected using the cold atomic absorption method at a wavelength of 253.7 nm in the detector's absorption cell.

The analytical procedure pertains to the indication of total mercury content (after converting the total mercury content into an atomic form). Mercury content is determined in lyophilized muscle tissue of great cormorants.

During the analytical procedure, the following reagents are used:

- Mercury standard—MSHG—100 ppm, concentration 100.48 ± 0.22 µg/mL in 3.3% HCl, Inorganic Ventures, Inc., USA
- L-Cysteine, 98%, Nacalai Tesque, Inc., Kyoto, Japan
- Additive B (activated alumina), Wako Pure Chemical Industries, Ltd., Japan
- Additive M (sodium carbonate and calcium hydroxide), POCh, Poland
- Nitric acid—suprapure, Merck, Germany
- Buffer solution pH 7.00 ± 0.05, POCh, Poland
- CRM: BCR-463: Total and methyl mercury in tuna fish, 2.85 ± 0.16 µg/g, IRMM, Geel, Belgium
- Deionized water

PREPARATION OF STANDARD SOLUTIONS

There are various methods available for preparing standard solutions. Nippon Instrument Corporation obtained good results using L-cysteine. However, in this case, the solution stability degrades with age or due to long storage in a warm place. Therefore, standard solutions should be kept in a cool, dark place.

PREPARATION OF 0.001% L-CYSTEINE SOLUTION

Measure 10 mg of L-cysteine and place it in a 1000 ml flask, then add water and 2 ml of guaranteed-reagent–grade concentrated nitric acid.

While ensuring uniformity of the contents in the flask by shaking it well, bring the total volume to 1000 ml by adding deionized water. For storage, keep in a cool and dark place.

STANDARD SOLUTION PREPARATION

Take 1 ml of 100 ppm solution and dilute it to 10 ml with 0.001% L-cysteine solution. Now a standard solution of 10-ppm has been prepared. By diluting in a similar manner, a standard solution of any concentration may be prepared. It should be noted that any mercury present in reagents or redistilled water should also be taken into consideration when a very diluted solution is prepared.

Any diluted solution, 100 ppm standard solution, and 10 ppm or less standard solution should be reprepared after 1 year or 6 months have elapsed, respectively.

Before using a new volumetric flask, wash it with acid. In particular, when any solution of 1 ppm or less is prepared, carefully wash the flask with acid and ensure that its tap is thoroughly washed.

It is acceptable to use commercially available undiluted standard stock solutions (100 ppm or 1000 ppm) of mercury intended for atomic absorptiometry as $HgCl_2$. However, ensure that any mercury contained is in the form of $HgCl_2$. Some products contain $Hg(NO_3)_2$ as a mercury component. Since $Hg(NO_3)_2$ may react with L-cysteine and lose its function as a fixing agent, do not use standard undiluted $Hg(NO_3)_2$ solutions.

Mercury has toxic properties; therefore, during the preparation of standard solutions, it is advisable to adhere to procedure guidelines for these types of substances. The work should be conducted under a fume hood, using pipettes during the preparation of standard solutions. Protective attire should be worn: safety glasses, rubber gloves, and a lab coat.

Care should also be taken while working with the atomic absorption analyzer, because of the high temperatures of some of its components, such as ovens heated up to 850°C.

For determining total mercury content in analyzed samples, an automatic mercury analyzer is utilized, MA-2000 from NIC (Japan). The Mercury/MA-2000 is a mercury analysis system that can measure mercury in liquid, solid, and gas (optional parts required) samples.

As shown in Figure 8.6, the system consists of the mercury analyzer (MA-2), the sample changer (BC-1), and a personal computer (hereafter referred to as PC). Once samples are in position in BC-1, each of them in turn is automatically transferred to the analyzer to be measured. The PC reads the resulting measurements

FIGURE 8.6   Mercury MA-2000 analysis system.

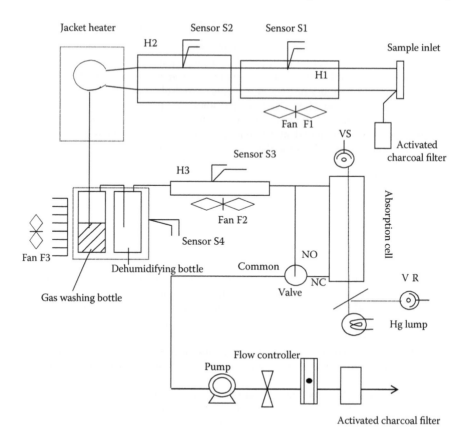

FIGURE 8.7   Schematic diagram of MA-2000.

in the order that the various analyses, including statistical calculations, can be performed.

A block diagram of the apparatus is presented in Figure 8.7.

ANALYTICAL PROCEDURE

Carefully separated bird tissues should be immediately deeply frozen, freeze-dried (lyophilized), and homogenized.

Homogenized samples should be stored in a refrigerator with a temperature of 0–6°C.

Homogenized samples should be directly weighed (10–50 ± 0.1 mg) into pre-cleaned combustion boats and automatically inserted into the Mercury/MA-2000 system (NIC—Japan).

To remove any interfering substances that are generated when thermally decomposing a sample, which would adversely affect measurements, gas washing is performed.

In addition, preheating the gold-coated diatomite particle support collection agent allows for the measurement to be done without the influence of any organic components, which would be physically absorbed to a certain extent, if not done so.

As a method of removing any substances that could interfere with the measurement, it is recommended that two kinds of additives be used: additive B (activated alumina) and additive M (sodium carbonate and calcium hydroxide). Before use, the additives should be subjected to a heat treatment in a heat treatment furnace at 750°C for at least three hours.

The sample boats that will be used should also be subjected to the same heat treatment.

The method for utilizing the additives is presented schematically in Figure 8.8.

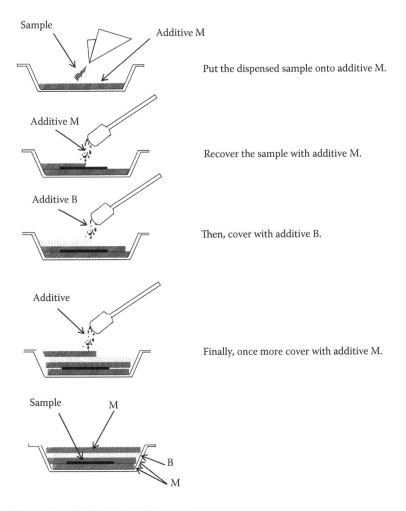

FIGURE 8.8   Method for using the additives.

CALIBRATION

Determine the calibration curve as a function of the peak surface area and the mercury content (Hg). Using an automatic pipette, dose at least five different volumes of the standard solution with a concentration of 1 ppm from the 20–100 μL section, which corresponds to 20–100 ng Hg.

For each mercury mass, repeat at least three times.

The minimal mass of the lyophilized tissue samples undergoing determination is limited on the one hand by the accuracy of the weight measurement, as well as the level of its homogeneity. Taking this into account, this value should not be less than 20 mg. However, the maximum sample mass is restricted by the maximum substance mass which can be introduced into the ceramic boat and consequently into the furnace. This value should not exceed 200 mg.

Taking this into account, the calibration curve corresponds to the range of Hg values in lyophilized tissue 0.1–5 ppm, or the values corresponding to the section of the values which most often appear in muscle tissue of great cormorants.

Draw the calibration curve and indicate the value of the regression parameters.

Compare these values with the determined values that are contained in the report.

The next steps of the analytical procedure are schematically presented in Figure 8.9.

During the analytical validation procedure, the values for the following parameters were indicated:

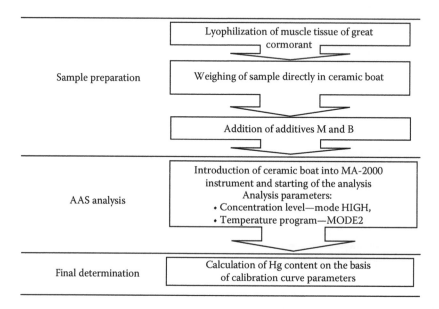

FIGURE 8.9 A schematic presentation of the analytical procedure for the determination of total mercury content in muscle tissue of great cormorant samples.

Selectivity

Applying the measurement technique ensures high selectivity for indicating mercury for two reasons:

1. The amalgamation reaction is a selective reaction for mercury.
2. The absorption radiation measurement is realized for mercury's characteristic wavelength.

Linearity

A series of standard solutions was prepared with a mercury content of 20 to 100 ng. For each of the solutions, three independent measurements were conducted, and based on the obtained results, regression parameters were indicated and the calibration curve was determined. The obtained values are presented in Table 8.6 and the calibration curve is presented in Figure 8.10.

A high regression coefficient r after fulfilling conditions for a "uniform" concentration distribution in terms of the calibration curve commands a high linear procedure.

TABLE 8.6

Calculated Regression Parameters for Linearity Determination

Number of results—n	25
Slope—b	1.730
Intercept—a	−2.1
Residual standard deviation—SD_{xy}	2.7
Standard deviation of the slope—SD_b	0.019
Standard deviation of the intercept—SD_a	1.3
Regression coefficient—r	0.9986

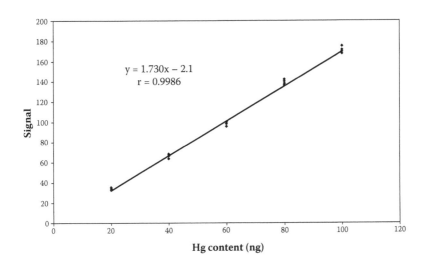

FIGURE 8.10   Calibration curve for linearity determination.

LIMIT OF DETECTION AND QUANTITATION

The *LOD* value is determined based on a series of measurement results for standard solution samples with the three lowest levels of mercury content (20, 40, and 60 ng). A calibration curve was outlined based on the obtained measurement results, parameters which determined *LOD* values, and the relationship:

$$LOD = \frac{3.3 \times SD}{b}$$

A calibration plot is presented in Figure 8.11.

The *LOD* value was deemed to be 2.5 ng, which, assuming the sample mass which underwent indication of an even 20 mg, corresponds to the mercury concentration in tissue samples of an even 0.12 ppm. However, the *LOQ* value was determined to be *LOQ* = 3·*LOD*, equaling 7.4 ng (assuming the mass of the 20 mg sample corresponds to a concentration of 0.37 ppm).

RANGE

The measurement range is a concentration range from the *LOQ* section, to a maximum standard solution concentration used for calibration. Therefore, it is equal to

$$7.4 \div 100 \text{ [ng]}$$

which, assuming the mass of the sample which underwent indication is an even 20 mg, corresponds to a mercury concentration of

$$0.37 \div 5.0 \text{ [ppm]}$$

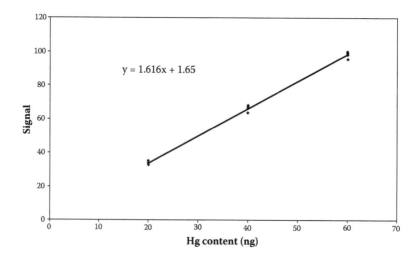

FIGURE 8.11   Calibration curve for *LOD* determination.

REPEATABILITY

Repeatability is determined based on a series of measurement results for three real samples (muscle tissue of great cormorant after lyophilization). This value is determined as an average CV value for three series.

The determined repeatability value is equal to $CV_{repeatability}$—4.24%.

TRUENESS

The trueness value is determined based on determination results for certified reference material samples (BCR–463—Tuna fish: total Hg and methylmercury) and is presented as a recovery value. A series of 5 independent determinations is conducted. The determined trueness value is equal to 97.1 ± 8.2%.

The determined trueness value is graphically presented in Figure 8.12.

UNCERTAINTY

The main components of the uncertainty budget were the uncertainty value resulting from the determination of the calibration curve, the uncertainty value related to the unrepeatability of measurement results, as well as the uncertainty value indicating trueness.

An estimation of the combined uncertainty value is conducted using the calculation

$$u_{smpl} = \sqrt{u_{cal}^2 + u_{rep}^2 + u_{true}^2}$$

where

u_{smpl}: Combined relative standard uncertainty for determined results for the real sample

u_{cal}: Relative standard uncertainty related to the calibration step

u_{rep}: Relative standard uncertainty related to repeatability of measurement results

u_{true}: Relative standard uncertainty related to indicating trueness

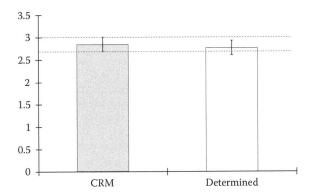

FIGURE 8.12 Comparison of the determined value with a certified Hg content value—trueness determination.

Determination of standard uncertainty related to the calibration step (preparation of a series of standard solutions, realization of measurements for the series of standard solutions, an approximation of measurement points of the calibration line with the use of linear regression) is conducted based on calibration parameters. Calculations are conducted for minimal masses for each of the analyzed real samples.

The calculated uncertainty value for $k = 2$ equals 9.3% (as an average of the three samples).

During the revalidation process, attention should be paid to the stability of the calibration curve. The determined value parameters for the calibration curve should not differ by more than ± 5% in relation to values determined during the validation process (Table 8.6.).

The consecutive parameter is trueness, indicated based on CRM determinations, as well as repeatability, whose value should not exceed $CV = 5\%$.

CONCLUSIONS

This analytical procedure fulfils requirements for a procedure serving to determine whole mercury content in lyophilized tissue samples from muscle tissue of great cormorants.

The procedure is characterized by high selectivity, repeatability ($CV = 4.24\%$), trueness (recovery = 97.1% ± 8.2%), and therefore, high precision.

Results obtained using this method are characterized by low uncertainty (about 10%).

The estimated limit of detection value $LOD = 2.5$ ng of total mercury in the sample, assuming the minimal mass is an even 20 mg, corresponds to a concentration of 0.12 ppm, and allows for the discovery of trace amounts of mercury in analyzed samples.

BIBLIOGRAPHY

Ambrus A., Quality Control of Pesticide Products, http://www-pub.iaea.org/MTCD/publications/PDF/te_1612_web.pdf, 2009 (access date 20.07.2017).

Analytical Detection Limit Guidance, Wisconsin Department of Natural Resources, 1996.

Analytical Methods Committee, Measurement of near zero concentration: Recording and reporting results that fall close to or below the detection limit, *Analyst*, 126, 256–259, 2001.

Boening D.W., Ecological effects, transport and fate of Mercury: A general review, *Chemosphere*, 40, 1335–1351, 2000.

Downs S.G., Macleod C.L., and Lester J.N., Mercury in precipitation and its relation to bioaccumulation in fish: A literature review, *Water Air Soil Pollut.*, 108, 149–187, 1998.

EURACHEM Guide: The Fitness for Purpose of Analytical Methods, A Laboratory Guide to Method Validation and Related Topics, Second Internet Edition, 2014, https://www.eurachem.org/images/stories/Guides/pdf/MV_guide_2nd_ed_EN.pdf (access date 20.07.2017).

Houserová P., Hedbavny J., Matejcek D., Kracmar S., Sitko J., and Kuban V., Determination of total mercury in muscle, intestines, liver and kidney tissues of cormorant (Phalacrocorax carbo), great crested grebe (Podiceps cristatus) and eurasian buzzard (Buteo buteo), *Vet. Med.—Czech*, 50 (2), 61–68, 2005.

Houserová P., Kubáň V., Komar S., and Sitko J., Total Mercury and Mercury species in birds and fish in an aquatic ecosystem in the Czech Republic, *Environ. Pollut.* 145, 185–194, 2007.

Huber L., Validation of Analytical Methods, http://www.chem.agilent.com/Library/primers /Public/5990-5140EN.pdf, 2010 (access date 20.07.2017).

International Conference on Harmonization (ICH) of Technical Requirements for the Registration of Pharmaceuticals for Human Use: Text on Validation of Analytical Procedures, ICH-Q2A, Geneva, 1994.

International Conference on Harmonization (ICH) of Technical Requirements for the Registration of Pharmaceuticals for Human Use: Validation of Analytical Procedures: Metrology, ICH-Q2B, Geneva, 1996.

International Vocabulary of Metrology—Basic and general concepts and associated terms (VIM), Joint Committee for Guides in Metrology, JCGM 200:2012.

Johnston R.K., and Valente R.M., Specifying and Evaluating Analytical Chemistry Quality Requirements for Ecological Risk Assessment, Marine Environmental Support Office, Technical Memorandum 99-01, San Diego, 2000.

Kim E.Y., Saeki K., Tanabe S., Tanaka H., and Tatsukawa R., Specific accumulation of mercury and selenium in seabirds, *Environ. Pollut.*, 94, 261–265, 1996.

Konieczka P., Misztal-Szkudlińska M., Namieśnik J., and Szefer P., Determination of total mercury in fish and cormorant using cold vapour atomic absorption spectrometry, *Pol. J. Environ. Stud.*, 19, 931–936, 2010.

Mercury Analysis System, "Mercury/MA-2000," Instruction Manual, Nippon Intruments Corporation, No. NIC-600-2009-04.

Misztal-Szkudlińska M., Szefer P., Konieczka P., and Namieśnik J., Biomagnification of mercury in trophic relation of Great Cormorant (Phalacrocorax carbo) and fish in the Vistula Lagoon, Poland, *Environ. Monit. Assess.*, 176, 439–449, 2011.

Nam D.H., Anan Y., Ikemoto T., Okabe Y., Kim E.-Y, Subramanian A., Saeki K., and Tanabe S., Specific accumulation of 20 trace elements in great cormorants (Phalacrocorax carbo) from Japan, *Environ. Pollut.* 134, 503–514, 2005.

Saeki K., Okabe Y., Kim E.-Y., Tanabe S., Fukuda M., and Tatsukawa R., Mercury and cadmium in common cormorants (Phalacrocorax carbo), *Environ. Pollut.* 108, 249–255, 2000.

Thompson M., Ellison S.L.R., and Wood R., Harmonized guidelines for single-laboratory validation of methods of analysis, *Pure Appl. Chem.*, 74, 835–855, 2002.

United States Pharmacopeial Convention, United States Pharmacopeia 23, US Rockville, 1995.

Vogelgesang J., and Hädrich J., Limits of detection, identification and determination: a statistical approach for practitioners, *Accred. Qual. Assur.*, 6, 242–255, 1998.

The validation process was conducted by Dr. Piotr Konieczka
The validation report was checked and confirmed by Prof. Jacek Namieśnik
Gdańsk, 20 June 2008

REFERENCES

1. Danzer K., A closer look at analytical signals, *Anal. Bioanal. Chem.*, 380, 376–382, 2004.

2. Huber L., Validation of analytical methods, http://www.chem.agilent.com/Library primers/Public/5990-5140EN.pdf, 2010 (access date July 20, 2017).

3. Ambrus A., Quality control of pesticide products, http://www-pub.iaea.org/MTCD /publications/PDF/te_1612_web.pdf , 2009 (access date July 20, 2017).

4. Konieczka P., The role of and place of method validation in the quality assurance and quality control (QA/QC) system, *Crit. Rev. Anal. Chem.*, 37, 173–190, 2007.

5. Traverniers I., De Loose M., and Van Bockstaele E., Trends in quality in the analytical laboratory. II. Analytical method validation and quality assurance, *Trends Anal. Chem.*, 23, 535–552, 2004.

6. International Conference on Harmonization (ICH) of Technical Requirements for the Registration of Pharmaceuticals for Human Use, Text on Validation of Analytical Procedures, ICH-Q2A, Geneva, 1994.

7. International Conference on Harmonization (ICH) of Technical Requirements for the Registration of Pharmaceuticals for Human Use, Validation of Analytical Procedures: Metrology, ICH-Q2B, Geneva, 1996.

8. United States Pharmacopeial Convention, United States Pharmacopeia 23, US Rockville, 1995.

9. Konieczka P., and Namieśnik J., eds., Kontrola i zapewnienie jakości wyników pomiarów analitycznych, WNT, Warsaw, 2007 (in Polish).

10. Vesseman J., Stefan R.I., Van Staden J. F., Danzer K., Lindner W., Burns D.T., Fajgelj A., and Müller H., Selectivity in analytical chemistry (IUPAC Recommendations 2001), *Pure Appl. Chem.*, 73, 1381–1386, 2001.

11. Valcárcel M., Gómez-Hens A., and Rubio S., Selectivity in analytical chemistry revisited, *Trends Anal. Chem.*, 20, 386–393, 2001.

12. Kapeller R., Quantifying selectivity: A statistical approach for chromatography, *Anal. Bioanal. Chem.*, 377, 1060–1070, 2003.

13. Danzer K., and Currie L.A., Guidelines for calibration in analytical chemistry, *Pure Appl. Chem.*, 70, 993–1014, 1998.

14. Thompson M., Ellison S.L.R., and Wood R., Harmonized guidelines for single-laboratory validation of methods of analysis, *Pure Appl. Chem.*, 74, 835–855, 2002.

15. Dobecki M. (ed.), Zapewnienie jakości analiz chemicznych, Instytut Medycyny Pracy im. Prof. J. Nofera, Łódź, 2004 (in Polish).

16. Ellison S.L.R., In defense of the correlation coefficient, *Accred. Qual. Assur.*, 11, 146–152, 2006.

17. González A.G., Herrador M.Á., Asuero A.G., and Sayago A., The correlation coefficient attacks again, *Accred. Qual. Assur.*, 11, 256–258, 2006.

18. Asuero A.G., Sayago A., and González A.G., The correlation coefficient: An overview, *Crit. Rev. Anal. Chem.*, 36, 41–59, 2006.

19. Van Loco J., Elskens M., Croux C., and Beernaert H., Linearity of calibration curves: Use and misuse of the correlation coefficient, *Accred. Qual. Assur.*, 7, 281–285, 2002.

20. Hibbert D.B., Further comments on the (miss-)use of *r* for testing the linearity of calibration functions, *Accred. Qual. Assur.*, 10, 300–301, 2004.

21. Huber W., On the use of the correlation coefficient *r* for testing the linearity of calibration functions, *Accred. Qual. Assur.*, 9, 726–726, 2004.

22. De Souza S.V.C., and Junqueira R.G., A procedure to assess linearity by ordinary least squares method, *Anal. Chim. Acta*, 552, 25–35, 2005.

23. Mulholland M., and Hibbert D.B., Linearity and the limitations of least squares calibration, *J. Chromatogr. A*, 762, 73–82, 1997.

24. Michulec M., and Wardencki W., Development of headspace solid-phase microextraction-gas chromatography method for the determination of solvent residues in edible oils and pharmaceuticals, *J. Chromatogr. A*, 1071, 119–124, 2005.

25. Michulec M., and Wardencki W., The application of single drop extraction technique for chromatographic determination of solvent residues in edible oils and pharmaceutical products, *Chromatographia*, 64, 191–197, 2006.

26. International vocabulary of metrology—Basic and general concepts and associated terms (VIM), Joint Committee for Guides in Metrology, JCGM 200:2012.

27. Konieczka P., Sposoby wyznaczania granicy wykrywalności i oznaczalności, *Chem. Inż. Ekol.*, 10, 639–654, 2003 (in Polish).

28. EURACHEM Guide: The Fitness for Purpose of Analytical Methods, A Laboratory Guide to Method Validation and Related Topics, Second Internet Edition, 2014, https://www.eurachem.org/images/stories/Guides/pdf/MV_guide_2nd_ed_EN.pdf (access date July 20, 2017).

29. Analytical Detection Limit Guidance, Wisconsin Department of Natural Resources, 1996.

30. Geiß S., and Einax J.W., Comparison of detection limits in environmental analysis—is it possible? An approach on quality assurance in the lower working range by verification, *Fresenius J. Anal. Chem.*, 370, 673–678, 2001.

31. Namieśnik J., and Górecki T., Quality of analytical results, *Rev. Roum. Chim.*, 46, 953–962, 2001.

32. Świtaj-Zawadka A., Konieczka P., Przyk E., and Namieśnik J., Calibration in metrological approach, *Anal. Lett.*, 38, 353–376, 2005.

33. Accuracy (trueness and precision) of measurement methods and results—Part 1: General principles and definitions, ISO 5725-1, 1994.

34. Accuracy (trueness and precision) of measurement methods and results—Part 2: Basic method for the determination of repeatability and reproducibility of a standard measurement method, ISO 5725-2, 1994.

35. Accuracy (trueness and precision) of measurement methods and results—Part 3: Intermediate measures of the precision of standard measurement method, ISO 5725-3, 1994.

36. Accuracy (trueness and precision) of measurement methods and results—Part 5: Alternative methods for the determination of the precision of a standard measurement method, ISO 5725-5, 1994.

37. De Castro M., Bolfarine H., Galea-Rojas M., and De Castilho M.V., An exact test for analytical bias detection, *Anal. Chim. Acta*, 538, 375–381, 2005.

38. Accuracy (trueness and precision) of measurement methods and results—Part 4: Basic method for the determination of the trueness of a standard measurement method, ISO 5725-4, 1994.

39. Accuracy (trueness and precision) of measurement methods and results—Part 6: Use in practice accuracy values, ISO 5725-6, 1994.

40. Hulanicki A., Absolute methods in analytical chemistry, *Pure Appl. Chem.*, 67(11), 1905–1911, 1995.

41. Dybczyński R., Danko B., Polkowska-Motrenko H., and Samczyński Z., RNAA in metrology: A higly accurate (definitive) method, *Talanta*, 71, 529–536, 2001.

42. Richter W., Primary methods of measurement in chemical analysis, *Accred. Qual. Assur.*, 2, 354–359, 1997.

43. Hibbert D.B., Systematic errors in analytical chemistry, *J. Chromatogr. A.*, 1158, 25–32, 2007.

44. Hulanicki A., Współczesna chemia analityczna, PWN, Warsaw, 2001 (in Polish).

45. Kozłowski E., Statystyczne kryteria oceny wyników i metod analitycznych, in Bobrański B. (ed.), *Analiza ilościowa związków organicznych*, PWN, Warsaw, 1979 (in Polish).

46. Cuadros-Rodríguez L., Romero R., and Bosque-Sendra J.M., The role of the robustness/ruggedness and inertia studies in research and development of analytical processes, *Crit. Rev. Anal. Chem.*, 35, 57–69, 2005.

47. Dejaegher B., and Vander Heyden Y., Ruggedness and robustness testing, *J. Chromatogr. A.*, 1158, 138–157, 2007.

9 Method Equivalence

9.1 INTRODUCTION

Method equivalence is defined as a measurement method other than the reference method for the measurement for which equivalence has been demonstrated.

In cases where it is not possible to use the reference method (norm) in the laboratory, for example, due to the lack of a suitable apparatus, it is necessary to document method equivalence. This is confirmation that the results obtained by the method used in the laboratory agree with the reference method. Method equivalence shall also apply in cases where the norm method is more expensive and time-consuming than that which is used in the laboratory. Method equivalence is the answer to the question of whether the parameters of the test methods and reference methods are significantly different or statistically significant. This is particularly required in the case of the nonregulated method, to prove there are no statistically significant differences in the results, for example, in the course of the accreditation process.

9.2 WAYS OF EQUIVALENCE DEMONSTRATION

Validation parameters, as described in Chapter 8, are calculated based on the values of the statistical parameters such as mean (trueness, accuracy) or the standard deviation (linearity, *LOD*, *LOQ*, precision, robustness, ruggedness).

For this reason, demonstration of method equivalence is first and foremost an indication of compliance obtained using the exanimated method and the reference method values of mean and standard deviation. Depending on the type of data sets, strategy of the equivalence method, demonstration, there are three basic ways of proceeding.

9.2.1 DIFFERENCE TESTING [1–4]

Difference tests have been widely used to answer questions about whether a disparity has been successfully addressed; however, these tests are subject to well-known limitations, and the results are sometimes misinterpreted. In tests of difference, analysts test the null hypothesis that the set of data under consideration do not differ. In difference testing, the null hypothesis is "no difference." If the analysis reveals a statistically significant difference between groups, the null hypothesis of no difference is rejected. However, if the analysis does not reveal a statistically significant difference between groups, the null hypothesis must stand—it cannot be rejected.

As statistical tests, Student's t (for mean comparison) and Snedecor's F (for standard deviation comparison) are mainly used.

For the comparison of more than two sets of data, ANOVA is often used.

Using a difference test is the answer to the question: Is it likely that no difference exists between two sets of results?

Example 9.1

Problem: Determine the equivalence of the examined method based on the given series of measurement results, results of determination for CRM, and the precision of reference method. Assume the value $\alpha = 0.05$.

Data: Result series, mg/dm³:

	Data
1	12.56
2	12.75
3	13.11
4	12.31
5	12.98
6	13.06

CRM, mg/dm³:

x_{CRM}	10.56	x_{det}	11.6
U_{CRM}	0.65	U_{det}	1.5
k	2	k	2

Precision of reference method: $CV_o = 2.0\%$

Solution:

1. Check for outliers, using Dixon's Q test:

No. of results—n	6
Range—R	0.80
Q_1	0.313
Q_n	0.062
Q_{crit}	0.560

According to the equation from Section 1.8.3.

Because Q_1 and $Q_n < Q_{crit}$, there is no outlier in the results series. The calculated values of x_m, SD and CV are

x_m	12.80	mg/dm³
SD	0.32	mg/dm³
CV	2.5	%

2. Check (at the level of significance $\alpha = 0.05$) if the calculated CV differs statistically significantly from the CV_o for reference method. Apply the $\chi2$ test.

Number of results—n	6
CV	2.5
χ^2	9.09
χ^2_{crit} $(f = 5, \alpha = 0.05)$	11.07

According to the equation from Section 1.8.4.

Because $\chi^2 < \chi^2_{crit}$, there is not a statistically significant difference in CV values (precision). The examined method does not differ statistically significantly in precision.

3. Compare the result obtained for CRM with certified value; calculate trueness as a recovery value for $k = 2$.

%R	110%
$U(k=2)_{\%R}$	15%

$$\%R = \frac{x_{det}}{x_{CRM}}[\%] \qquad U = k \cdot \frac{\sqrt{\left(u^2_{(x_{det})} + u^2_{(x_{CRM})}\right)}}{\left(\dfrac{x_{det} + x_{CRM}}{2}\right)}$$

A value of 100% is in the range of calculated trueness value.

Conclusion: The results obtained by the investigated method do not differ statistically significantly from results obtained by the reference method.
Excel file: exampl_equivalence01.xls

Example 9.2

Problem: Determine the equivalence of the examined method based on the given series of measurement results, obtained by examined method and reference method.
Data: Result series, mg/dm³:

	Data	
	Examined Method	**Reference Method**
1	12.56	13.07
2	12.75	13.23
3	13.11	13.10
4	12.31	12.98
5	12.98	13.33
6	13.06	13.06

Solution:

1. Check for outliers using the Dixon's Q test:

	Examined Method	Reference Method
No. of results—n	6	6
Range—R	0.80	0.35
Q_1	0.313	0.229
Q_n	0.062	0.286
Q_{crit}	0.560	0.560

According to the equation from Section 1.8.3.
Because Q_1 and $Q_n < Q_{crit}$, for both series, there are no outliers in the results series. The calculated values of x_m, SD, and CV are

	Examined Method	Reference Method	
x_m	12.80	13.13	mg/dm³
SD	0.32	0.13	mg/dm³
CV	2.5	1.0	%

2. Check (at the level of significance $\alpha = 0.05$) if the standard deviation values for both the series are statistically significantly different.
   Apply Snedecor's F test:

F	6.06
F_{crit}	5.05

According to the equation from Section 1.8.5.
Because $F > F_{crit}$, there is a statistically significant difference in variance values for the compared series, the series differ in precision.

Conclusion: The results obtained by the investigated method differ statistically significantly from results obtained by the reference method.
Excel file: exampl_equivalence02.xls

Example 9.3

Problem: Determine the equivalence of the examined method based on the given series of measurement results, obtained by the examined method and reference method.

Data: Result series, mg/dm^3:

	Examined Method	Reference Method
	Data	
	Examined Method	Reference Method
1	12.56	13.07
2	12.75	13.27
3	13.11	13.10
4	12.31	12.91
5	12.98	13.33
6	13.06	13.06

Solution:

1. Check for outliers using Dixon's Q test:

	Examined Method	Reference Method
No of results—n	6	6
Range—R	0.80	0.42
Q_1	0.313	0.357
Q_n	0.062	0.143
Q_{crit}	0.560	0.560

According to the equation from Section 1.8.3.
Because Q_1 and $Q_n < Q_{crit}$, for both series, there are no outlier in the results series. The calculated values of xm, SD and CV are

	Examined method	Reference method	
x_m	12.80	13.12	mg/dm^3
SD	0.32	0.15	mg/dm^3
CV	2.5	1.2	%

2. Check (at the level of significance $\alpha = 0.05$) if the standard deviation values for both the series are statistically significantly different.
   Apply Snedecor's F test:

F	4.24
F_{crit}	5.05

According to the equation from Section 1.8.5.
Because $F < F_{crit}$, there is no statistically significant difference in variance values for the compared series, and the series do not differ in precision.

3. Check (at the level of significance $\alpha = 0.05$) if the means for both the
series are statistically significantly different.
   Apply Student's t test:

t	2.296
t_{crit}	2.228

According to the equation from Section 1.8.9.
   Because $t > t_{crit}$, there is a statistically significant difference in means
for the compared series, and the series differ in accuracy.

Conclusion: The results obtained by the investigated method differ statistically
significantly from results obtained by the reference method.
Excel file: exampl_equivalence03.xls

Example 9.4

Problem: Determine the equivalence of the examined method based on the given
series of measurement results, obtained by the examined method and reference
method.
Data: Result series, mg/dm^3:

	Data	
	Examined Method	**Reference Method**
1	12.56	13.07
2	12.75	13.27
3	13.11	13.10
4	12.31	12.91
5	12.98	13.74
6	13.06	13.06

Solution:

1. Check for outliers using Dixon's Q test:

	Examined Method	**Reference Method**
No. of results—n	6	6
Range—R	0.80	0.83
Q_1	0.313	0.181
Q_n	0.062	0.566
Q_{crit}	0.560	0.560

According to the equation from Section 1.8.3.

Because $Q_n > Q_{crit}$, for reference method series, there is an outlier in the series. So the values of x_m, SD, and CV were calculated for 6 results in the examined method series and 5 results in the reference method:

	Examined Method	Reference Method	
x_m	12.80	13.08	mg/dm³
SD	0.32	0.13	mg/dm³
CV	2.5	1.0	%
n	6	5	

2. Check (at the level of significance $\alpha = 0.05$) if the standard deviation values for both the series are statistically significantly different.
   Apply Snedecor's F test:

F	6.02
F_{crit}	6.26

According to the equation from Section 1.8.5.

Because $F < F_{crit}$, there is no statistically significant difference in variance values for the compared series, so the series do not differ in precision.

3. Checking (at the level of significance $\alpha = 0.05$) if the means for both the series are statistically significantly different.
   Apply Student's t test:

t	1.897
t_{crit}	2.262

According to the equation from Section 1.8.9.

Because $t < t_{crit}$, there is no statistically significant difference in means for the compared series, and the series do not differ in accuracy.

Conclusion: The results obtained by the investigated method do not differ statistically significantly from the results obtained by the reference method.
Excel file: exampl_equivalence04.xls

9.2.2 EQUIVALENCE TESTING [1–6]

In equivalence testing, the null hypothesis is formulated so that the statistical test is proof of similarity; it states that the groups differ by more than a tolerably small amount.

In equivalence testing, the null hypothesis is "a difference of certain limit or more." In equivalence testing, the null hypothesis states that the difference among group means is greater than some minimal difference representing practical equivalence. The alternative hypothesis is that the difference is not greater than this specified minimum difference.

Equivalence testing is used when one wants assurance that the means do not differ by too much. In other words, the means are practically equivalent. A threshold difference acceptance criteria is set by the analyst for each parameter under test. The means are considered equivalent if the difference in the two groups is significantly lower than the upper practical limit and significantly higher than the lower practical limit.

So equivalence tests can be used for studying the following:

- Comparison to a reference standard or target
- Comparison between two series
- Comparison of slopes for stability
- Comparison of intercepts

If one wants to determine equivalence, a more appropriate statistical question to ask is perhaps: Is there an unacceptable difference between the two sets of results?

If Student's t test has to be applied for that purpose, the following modified equation has to be used:

$$t = \frac{\left| x_{m_{ref}} - x_{m_{exam}} \right| - \Delta\% \cdot x_{m_{ref}}}{\sqrt{(n_1 - 1)SD_1^2 + (n_2 - 1)SD_2^2}} \sqrt{\frac{n_1 n_2 (n_1 + n_2 - 2)}{n_1 + n_2}} \tag{9.1}$$

where
Δ [in%]: A limit of differences between compared values

Example 9.5

Problem: Determine the equivalence of the examined method based on the given series of measurement results, obtained by the examined method and reference method. In the case of means, comparison takes into account a limit of difference equal to ± 3%.
Data: Result series, mg/dm³:

	Data	
	Examined Method	Reference Method
1	12.56	13.07
2	12.75	13.27
3	13.11	13.45
4	12.31	13.14
5	12.98	13.3
6	13.06	13.06

Solution:

1. Check for outliers using Dixon's Q test:

	Examined Method	Reference Method
No of results—n	6	6
Range—R	0.80	0.39
Q_1	0.313	0.026
Q_n	0.062	0.308
Q_{crit}	0.560	0.560

According to the equation from Section 1.8.3.
Because Q_1 and $Q_n < Q_{crit}$, for both series, there are no outliers in the results series. The calculated values of x_m, SD and CV are

	Examined Method	Reference Method	
x_m	12.80	13.22	mg/dm³
SD	0.32	0.16	mg/dm³
CV	2.5	1.2	%

2. Check (at the level of significance $\alpha = 0.05$) if the standard deviation values for both the series are statistically significantly different. Apply Snedecor's F test:

F	4.07
F_{crit}	5.05

According to equation from Section 1.8.5.
Because $F < F_{crit}$, there is no statistically significant difference in the variance values for the compared series; so the series do not differ in precision.
3. Check (at the level of significance $\alpha = 0.05$) if the means for both the series are statistically significantly different. Take into account a limit of difference equal to $\pm 3\%$.
Apply Student's t test:

$$t = \frac{\left| x_{m_{ref}} - x_{m_{exam}} \right| - 3\% \cdot x_{m_{ref}}}{\sqrt{(n_1 - 1)SD_1^2 + (n_2 - 1)SD_2^2}} \sqrt{\frac{n_1 n_2 (n_1 + n_2 - 2)}{n_1 + n_2}}$$

t	0.198
t_{crit}	2.262

Because $t < t_{crit}$ there is no statistically significant difference in means for the compared series, so the series do not differ in accuracy.

Conclusion: The results obtained by the investigated method do not differ statistically significantly from the results obtained by the reference method.
Excel file: exampl_equivalence05.xls

9.2.3 REGRESSION ANALYSIS TESTING

In the case where it is possible to have sets of data for different content obtained by using both methods, it is recommended to apply regression analysis. The way to proceed is then to calculate regression line parameters and, using Student's t test, compare an obtained values with expected ones.

Example 9.6

Problem: Determine the equivalence of the examined method based on the given series of measurement results for real samples, obtained by the examined method and reference method. Apply the linear regression method.
Data: Result series, ppb:

	Examined Method	Reference Method
	y	x
1	46.9	45.7
2	88.5	86.9
3	101.0	97.8
4	79.4	77.2
5	21.2	19.6
6	12.3	10.9
7	109.0	105.0
8	59.3	56.8
9	57.3	56.2
10	47.2	44.2
11	39.3	35.2
12	38.1	37.2
13	27.3	26.8
14	90.2	89.3
15	111.0	106.0

Solution:
Using the linear regression method, calculate regression parameters:

No. of results—n	15	
b	1.021	
a	0.95	ppb
SD_{xy}	1.24	
SD_a	0.70	ppb
SD_b	0.010	
r	0.999	

Check the statistically significant differences among parameter b and 1 and parameter a and 0, and apply Student's t test.

$$t_b = \frac{|b-1|}{SD_b} \qquad\qquad t_a = \frac{|a|}{SD_a}$$

t_b	2.041
t_a	1.364
t_{crit}	2.160

Because $t_b < t_{crit}$ and $t_a < t_{crit}$, there is no statistically significant difference in results obtained by both methods.
Conclusion: The results obtained by the investigated method do not differ statistically significantly from the results obtained by the reference method.
Excel file: exampl_equivalence06.xls

Example 9.7

Problem: Determine the equivalence of the examined method based on the given series of measurement results for real samples, obtained by the examined method and reference method. Apply the linear regression method.
Data: Result series, g/l:

	Examined Method	Reference Method
	y	x
1	12.3	11.6
2	9.4	8.8
3	3.2	2.7
4	15.8	14.9
5	17.4	15.9
6	21.0	19.3
7	21.3	19.6
8	33.8	30.1

Solution:
Using the linear regression method, calculate the following regression parameters:

No of results—n	8	
b	1.119	
a	−0.42	g/l
SD_{xy}	0.41	
SD_a	0.32	g/l
SD_b	0.019	
r	0.999	

Check the statistically significant differences among parameter b and 1 and parameter a and 0, and apply Student's t test.

$$t_b = \frac{|b-1|}{SD_b} \qquad t_a = \frac{|a|}{SD_a}$$

t_b	6.330
t_a	1.299
t_{crit}	2.447

Because $t_b > t_{crit}$ and $t_a < t_{crit}$, there is a statistically significant difference in results obtained by both methods.

Conclusion: The results obtained by the investigated method differ statistically significantly from the results obtained by the reference method.

Excel file: exampl_equivalence07.xls

9.3   CONCLUSION

Changes and differences in analytical methods may cause significant changes in the obtained results. A comparison of two methods (and in fact their metrological parameters) can demonstrate their equivalence or lack thereof. In such cases, there is a need to assess the equivalence of the results achieved by the two methods. The equivalence test offers benefits compared to only checking the validation parameters, because the criteria to determine the correctness of a single method does not always mean the identity of the two independent methods, but their compliance [1].

REFERENCES

1. Chambers D., Kelly G., Limentani G., Lister A., Lung K.R., and Warner E., Analytical method equivalency an acceptable analytical practice, *Pharm. Techn.*, 64–80, September 2005.
2. Lung K.R., Gorko M.A., Llewelyn J., and Wiggins N., Statistical method for the determination of equivalence of automated test procedures, *J. Autom. Methods Manag. Chem.*, 25(6), 123–127, 2003.

3. Ermer J., Limberger M., Lis K., and Wätzig H., The transfer of analytical procedures, *J. Pharmaceut. Biomed.*, 85, 262–276, 2013.

4. Rogers J.L., Howard K.I., and Vessey J.T., Using significance tests to evaluate equivalence between two experimental groups, *Psychol. Bull.*, 113, 553–565, 1993.

5. Stegner B.L., Bostrom AG., Greenfield T.X., and Secombes C.J., Equivalence testing for use in psychosocial and services research: An introduction with examples, *Eval. Program Plann.*, 19(3), 193–198, 1996.

6. Stegner B.L., Bostrom AG., Greenfield T.X., and Secombes C.J., Equivalence testing for use in psychosocial and services research: An introduction with examples, *Eval. Program Plann.*, 19(4), 533–540, 1996.

Appendix

TABLE A.1
Critical Values, Student's *t* Test

f	$\alpha = 0.05$	$\alpha = 0.01$
1	12.706	63.567
2	4.303	9.925
3	3.182	5.841
4	2.776	4.604
5	2.571	4.032
6	2.447	3.707
7	2.365	3.499
8	2.306	3.355
9	2.262	3.250
10	2.228	3.169
11	2.201	3.106
12	2.179	3.055
13	2.160	3.012
14	2.149	2.977
15	2.131	2.947
16	2.120	2.921
17	2.110	2.898
18	2.101	2.878
19	2.093	2.861
20	2.086	2.845
22	2.074	2.819
24	2.064	2.797
26	2.056	2.779
28	2.048	2.763
30	2.042	2.750
35	2.030	2.716
40	2.021	2.706
45	2.014	2.690
50	2.009	2.678
60	2.000	2.660
70	1.994	2.648
80	1.990	2.639
100	1.984	2.626
∞	1.960	2.576

TABLE A.2

Critical Values of Parameter w_α

f	$\alpha = 0.05$	$\alpha = 0.01$
1	1.409	1.414
2	1.645	1.715
3	1.757	1.918
4	1.814	2.051
5	1.848	2.142
6	1.870	2.208
7	1.885	2.256
8	1.895	2.294
9	1.903	2.324
10	1.910	2.348
11	1.916	2.368
12	1.920	2.385
13	1.923	2.399
14	1.926	2.412
15	1.928	2.423
16	1.931	2.432
17	1.933	2.440
18	1.935	2.447
19	1.936	2.454
20	1.937	2.460
22	1.940	2.470
24	1.941	2.479
26	1.943	2.487
28	1.944	2.492
30	1.945	2.498
35	1.948	2.509
40	1.949	2.518
45	1.950	2.524
50	1.951	2.529
60	1.953	2.537
70	1.954	2.542
80	1.955	2.547
100	1.956	2.553
∞	1.960	2.576

TABLE A.3
Critical Values of the z Parameter for Significance Level $\alpha = 0.05$

f	2	3	4	5	6	7	8	9	10	11	12
1	18.0	27.0	32.8	37.1	40.4	43.1	45.4	47.4	49.1	50.6	53.0
5	3.64	4.60	5.22	5.67	6.03	6.33	6.58	6.80	6.99	7.17	7.32
10	3.15	3.88	4.33	4.65	4.91	5.12	5.30	5.46	5.60	5.72	5.83
15	3.01	3.67	4.08	4.37	4.60	4.78	4.94	5.08	5.20	5.31	5.40
20	2.95	3.58	3.96	4.23	4.45	4.62	4.77	4.90	5.01	5.11	5.20
30	2.89	3.49	3.84	4.10	4.30	4.46	4.60	4.72	4.83	4.92	5.00
40	2.86	3.44	3.79	4.04	4.23	4.39	4.52	4.63	4.74	4.82	4.91
60	2.83	3.40	3.74	3.98	4.16	4.31	4.44	4.55	4.65	4.73	4.81
120	2.80	3.36	3.69	3.92	4.10	4.24	4.36	4.48	4.56	4.64	4.72
∞	2.77	3.31	3.63	3.86	4.03	4.17	4.29	4.39	4.47	4.55	4.62

The column group is headed by *n*.

TABLE A.4
Critical Values of Parameter z_α

f	$\alpha = 0.10$	$\alpha = 0.05$	$\alpha = 0.01$
2	2.06	2.46	3.23
3	1.71	1.96	2.43
4	1.57	1.76	2.14
5	1.50	1.66	1.98

TABLE A.5
Critical Values (Q_{crit}) of Dixon's Q Test

f	$\alpha = 0.10$	$\alpha = 0.05$	$\alpha = 0.01$
3	0.886	0.941	0.988
4	0.679	0.765	0.889
5	0.557	0.642	0.780
6	0.482	0.560	0.698
7	0.434	0.507	0.637
8	0.399	0.468	0.590
9	0.370	0.437	0.555
10	0.349	0.412	0.527

TABLE A.6

Critical Values (Q_{crit}) of Dixon's Q Test (Modification for $n \leq 40$)

f	$\alpha = 0.05$	$\alpha = 0.01$
3	0.970	0.994
4	0.829	0.926
5	0.710	0.821
6	0.628	0.740
7	0.569	0.680
8	0.608	0.717
9	0.564	0.672
10	0.530	0.635
11	0.502	0.605
12	0.479	0.579
13	0.611	0.697
14	0.586	0.670
15	0.565	0.647
16	0.546	0.627
17	0.529	0.610
18	0.514	0.594
19	0.501	0.580
20	0.489	0.567
21	0.478	0.555
22	0.468	0.544
23	0.459	0.535
24	0.451	0.526
25	0.443	0.517
26	0.436	0.510
27	0.429	0.502
28	0.423	0.495
29	0.417	0.489
30	0.412	0.483
31	0.407	0.477
32	0.402	0.472
33	0.397	0.467
34	0.393	0.462
35	0.388	0.458
36	0.384	0.454
37	0.381	0.450
38	0.377	0.446
39	0.374	0.442
40	0.371	0.438

TABLE A.7
Critical Values for χ^2 Test

f	$\alpha = 0.05$	$\alpha = 0.01$
1	3.84	6.64
2	5.99	9.21
3	7.81	11.34
4	9.49	13.28
5	11.07	15.09
6	12.59	16.81
7	14.07	18.48
8	15.51	20.09
9	16.92	21.67
10	18.31	23.21
11	19.68	24.72
12	21.03	26.22
13	22.36	27.69
14	23.68	29.14
15	25.00	30.58
16	26.30	32.00
17	27.59	33.41
18	28.87	34.80
19	30.14	36.19
20	31.41	37.57
21	32.67	38.93
22	33.92	40.29
23	35.17	41.64
24	36.41	42.98
25	37.65	44.31

TABLE A.8

Critical Values for Snedecor's _F_ test for Significance Level $\alpha = 0.05$ (Top Row) and $\alpha = 0.01$ (Bottom Row)

f_2					f_1					
	2	3	4	5	6	7	8	9	10	11
2	19.00	19.16	19.25	19.30	19.33	19.36	19.37	19.38	19.39	19.40
	99.01	99.17	99.25	99.30	99.33	99.34	99.36	99.38	99.40	99.41
3	9.55	9.28	9.12	9.01	8.94	8.88	8.84	8.81	8.78	8.76
	30.81	29.46	28.71	28.24	27.91	27.67	27.49	27.34	27.23	27.13
4	6.94	6.59	6.39	6.26	6.16	6.09	6.04	6.00	5.96	5.93
	18.00	16.69	15.98	15.52	15.21	14.98	14.80	14.66	14.54	14.45
5	5.79	5.41	5.19	5.05	4.95	4.88	4.82	4.78	4.74	4.70
	13.27	12.06	11.39	10.97	10.67	10.45	10.27	10.15	10.05	9.96
6	5.14	4.76	4.53	4.39	4.28	4.21	4.15	4.10	4.06	4.03
	10.92	9.78	9.15	8.57	8.47	8.26	8.10	7.98	7.87	7.79
7	4.74	4.35	4.12	3.97	3.87	3.79	3.73	3.68	3.63	3.60
	9.55	8.45	7.85	7.46	7.19	7.00	6.84	6.71	6.62	6.54
8	4.46	4.07	3.84	3.69	3.58	3.50	3.44	3.39	3.34	3.31
	8.65	7.59	7.01	6.63	6.37	6.19	6.03	5.91	5.82	5.74
9	4.26	3.86	3.63	3.48	3.37	3.29	3.23	3.18	3.13	3.10
	8.02	6.99	6.42	6.06	5.80	5.62	5.47	5.35	5.26	5.18
10	4.10	3.71	3.48	3.33	3.22	3.14	3.07	3.02	2.97	2.94
	7.56	6.55	5.99	5.64	5.39	5.21	5.06	4.95	4.85	4.78
11	3.98	3.59	3.36	3.20	3.09	3.01	2.95	2.90	2.86	2.82
	7.20	6.22	5.67	5.32	5.07	4.88	4.74	4.63	4.54	4.46

TABLE A.9

Critical Values, Hartley's F_{max} Test for Significance Level $\alpha = 0.05$

f					k					
	2	3	4	5	6	7	8	9	10	11
2	39.0	87.5	142	202	266	333	403	475	550	626
3	15.4	27.8	39.2	50.7	62.0	72.9	83.5	93.9	104	114
4	9.60	15.5	20.6	25.2	29.5	33.6	37.5	41.1	44.6	48.0
5	7.15	10.8	13.7	16.3	18.7	20.8	22.9	24.7	26.5	28.2
6	5.82	8.38	10.4	12.1	13.7	15.0	16.3	17.5	18.6	19.7
7	4.99	6.94	8.44	9.70	10.8	11.8	12.7	13.5	14.3	15.1
8	4.43	6.00	7.18	8.12	9.03	9.78	10.5	11.1	11.7	12.2
9	4.03	5.34	6.31	7.11	7.80	8.41	8.95	9.45	9.91	10.3
10	3.72	4.85	5.67	6.34	6.92	7.42	7.87	8.29	8.66	9.01
15	2.86	3.54	4.01	4.37	4.68	4.95	5.19	5.40	5.59	5.77
20	2.46	2.95	3.29	3.54	3.76	3.94	4.10	4.24	4.37	4.49
30	2.07	2.40	2.61	2.78	2.91	3.02	3.12	3.21	3.29	3.36
60	1.67	1.85	1.96	2.04	2.11	2.17	2.22	2.26	2.30	2.33
∞	1.00	1.00	1.00	1.00	1.00	1.00	1.00	1.00	1.00	1.00

TABLE A.10
Critical Values ν_o of the Aspin–Welch Test for Significance Level $\alpha = 0.05$

							c					
f_1	f_2	0.0	0.1	0.2	0.3	0.4	0.5	0.6	0.7	0.8	0.9	1.0
6	6	1.94	1.90	1.85	1.80	1.76	1.74	1.76	1.80	1.85	1.90	1.94
	8	1.94	1.90	1.85	1.80	1.76	1.73	1.74	1.76	1.79	1.82	1.86
	10	1.94	1.90	1.85	1.80	1.76	1.73	1.73	1.74	1.76	1.78	1.81
	15	1.94	1.90	1.85	1.80	1.76	1.73	1.71	1.71	1.72	1.73	1.75
	20	1.94	1.90	1.85	1.80	1.76	1.73	1.71	1.70	1.70	1.71	1.72
	∞	1.94	1.90	1.85	1.80	1.76	1.72	1.69	1.67	1.66	1.65	1.64
8	6	1.86	1.82	1.79	1.76	1.74	1.73	1.76	1.80	1.85	1.90	1.94
	8	1.86	1.82	1.79	1.76	1.73	1.73	1.73	1.76	1.79	1.82	1.86
	10	1.86	1.82	1.79	1.76	1.73	1.72	1.72	1.74	1.76	1.78	1.81
	15	1.86	1.82	1.79	1.76	1.73	1.71	1.71	1.71	1.72	1.73	1.75
	20	1.86	1.82	1.79	1.76	1.73	1.71	1.70	1.70	1.70	1.71	1.72
	∞	1.86	1.82	1.79	1.75	1.72	1.70	1.68	1.66	1.65	1.65	1.64
10	6	1.81	1.78	1.76	1.74	1.73	1.73	1.76	1.80	1.85	1.90	1.94
	8	1.81	1.78	1.76	1.74	1.72	1.72	1.73	1.76	1.79	1.82	1.86
	10	1.81	1.78	1.76	1.73	1.72	1.71	1.72	1.73	1.76	1.78	1.81
	15	1.81	1.78	1.76	1.73	1.72	1.70	1.70	1.71	1.72	1.73	1.75
	20	1.81	1.78	1.76	1.73	1.71	1.70	1.69	1.69	1.70	1.71	1.72
	∞	1.81	1.78	1.76	1.73	1.71	1.69	1.67	1.66	1.65	1.65	1.64
15	6	1.75	1.73	1.72	1.71	1.71	1.73	1.76	1.80	1.85	1.90	1.94
	8	1.75	1.73	1.72	1.71	1.71	1.71	1.73	1.76	1.79	1.82	1.86
	10	1.75	1.73	1.72	1.71	1.71	1.70	1.72	1.73	1.76	1.78	1.81
	15	1.75	1.73	1.72	1.70	1.70	1.69	1.70	1.70	1.72	1.73	1.75
	20	1.75	1.73	1.72	1.70	1.69	1.69	1.69	1.69	1.70	1.71	1.72
	∞	1.75	1.73	1.72	1.70	1.68	1.67	1.66	1.65	1.65	1.65	1.64
20	6	1.72	1.71	1.70	1.70	1.71	1.73	1.76	1.80	1.85	1.90	1.94
	8	1.72	1.71	1.70	1.70	1.70	1.71	1.73	1.76	1.79	1.82	1.86
	10	1.72	1.71	1.70	1.69	1.69	1.70	1.71	1.73	1.76	1.78	1.81
	15	1.72	1.71	1.70	1.69	1.69	1.69	1.69	1.70	1.72	1.73	1.75
	20	1.72	1.71	1.70	1.69	1.68	1.68	1.68	1.69	1.70	1.71	1.72
	∞	1.72	1.71	1.70	1.68	1.67	1.66	1.66	1.65	1.65	1.65	1.64
∞	6	1.64	1.65	1.66	1.67	1.69	1.72	1.76	1.80	1.85	1.90	1.94
	8	1.64	1.65	1.65	1.66	1.68	1.70	1.72	1.75	1.79	1.82	1.86
	10	1.64	1.65	1.65	1.66	1.67	1.69	1.71	1.73	1.76	1.78	1.81
	15	1.64	1.65	1.65	1.65	1.66	1.67	1.68	1.70	1.72	1.73	1.75
	20	1.64	1.65	1.65	1.65	1.66	1.66	1.67	1.68	1.70	1.71	1.72
	∞	1.64	1.64	1.64	1.64	1.64	1.64	1.64	1.64	1.64	1.64	1.64

TABLE A.11
Critical Values of Cochran's Test

p	n = 2		n = 3		n = 4		n = 5		n = 6	
	$\alpha = 0.01$	$\alpha = 0.05$	$\alpha = 0.01$	$\alpha = 0.05$	$\alpha = 0.01$	$\alpha = 0.05$	$\alpha = 0.01$	$\alpha = 0.05$	$\alpha = 0.01$	$\alpha = 0.05$
2	–	–	0.995	0.975	0.979	0.939	0.959	0.906	0.937	0.877
3	0.993	0.967	0.942	0.871	0.883	0.798	0.834	0.746	0.793	0.707
4	0.968	0.906	0.864	0.768	0.781	0.684	0.721	0.629	0.676	0.590
5	0.928	0.841	0.788	0.684	0.696	0.598	0.633	0.544	0.588	0.506
6	0.883	0.781	0.722	0.616	0.626	0.532	0.564	0.480	0.520	0.445
7	0.838	0.727	0.664	0.561	0.568	0.480	0.508	0.431	0.466	0.397
8	0.794	0.680	0.615	0.516	0.521	0.438	0.463	0.391	0.423	0.360
9	0.754	0.638	0.573	0.478	0.481	0.403	0.425	0.358	0.387	0.329
10	0.718	0.602	0.536	0.445	0.447	0.373	0.393	0.331	0.357	0.303
11	0.684	0.570	0.504	0.417	0.418	0.348	0.366	0.308	0.332	0.281
12	0.653	0.541	0.475	0.392	0.392	0.326	0.343	0.288	0.310	0.262
13	0.624	0.515	0.450	0.371	0.369	0.307	0.322	0.271	0.291	0.243
14	0.599	0.492	0.427	0.352	0.349	0.291	0.304	0.255	0.274	0.232
15	0.575	0.471	0.407	0.335	0.332	0.276	0.288	0.242	0.259	0.220
16	0.553	0.452	0.388	0.319	0.316	0.262	0.274	0.230	0.246	0.208
17	0.532	0.434	0.372	0.305	0.301	0.250	0.261	0.219	0.234	0.198
18	0.514	0.418	0.356	0.293	0.288	0.240	0.249	0.209	0.223	0.189
19	0.496	0.403	0.343	0.281	0.276	0.230	0.238	0.200	0.214	0.181
20	0.480	0.389	0.330	0.270	0.265	0.220	0.229	0.192	0.205	0.174
21	0.465	0.377	0.318	0.261	0.255	0.212	0.220	0.185	0.197	0.167
22	0.450	0.365	0.307	0.252	0.246	0.204	0.212	0.178	0.189	0.160
23	0.437	0.354	0.297	0.243	0.238	0.197	0.204	0.172	0.182	0.155
24	0.425	0.343	0.287	0.235	0.230	0.191	0.197	0.166	0.176	0.149
25	0.413	0.334	0.278	0.228	0.222	0.185	0.190	0.160	0.170	0.144
26	0.402	0.325	0.270	0.221	0.215	0.179	0.184	0.155	0.164	0.140
27	0.391	0.316	0.262	0.215	0.209	0.173	0.179	0.150	0.159	0.135
28	0.382	0.308	0.255	0.209	0.202	0.168	0.173	0.146	0.154	0.131
29	0.372	0.300	0.248	0.203	0.196	0.164	0.168	0.142	0.150	0.127
30	0.363	0.293	0.241	0.198	0.191	0.159	0.164	0.138	0.145	0.124
31	0.355	0.286	0.235	0.193	0.186	0.155	0.159	0.134	0.141	0.120
32	0.347	0.280	0.229	0.188	0.181	0.151	0.155	0.131	0.138	0.117
33	0.339	0.273	0.224	0.184	0.177	0.147	0.151	0.127	0.134	0.114
34	0.332	0.267	0.218	0.179	0.172	0.144	0.147	0.124	0.131	0.111
35	0.325	0.262	0.213	0.175	0.168	0.140	0.144	0.121	0.127	0.108
36	0.318	0.256	0.208	0.172	0.165	0.137	0.140	0.118	0.124	0.106
37	0.312	0.251	0.204	0.168	0.161	0.134	0.137	0.116	0.121	0.103
38	0.306	0.246	0.200	0.164	0.157	0.131	0.134	0.113	0.119	0.101
39	0.300	0.242	0.196	0.161	0.154	0.129	0.131	0.111	0.116	0.099
40	0.294	0.237	0.192	0.158	0.151	0.126	0.128	0.108	0.114	0.097

p: Number of laboratories.
n: Number of results for one level.

TABLE A.12
Critical Values of Grubbs' Test

p	One Greatest and One Smallest		Two Greatest and Two Smallest	
	Upper $\alpha = 0.01$	Lower $\alpha = 0.05$	Upper $\alpha = 0.01$	Lower $\alpha = 0.05$
3	1.155	1.155	–	–
4	1.496	1.481	0.0000	0.0002
5	1.764	1.715	0.0018	0.0090
6	1.973	1.887	0.0116	0.0349
7	2.139	2.020	0.0308	0.0708
8	2.274	2.126	0.0563	0.1101
9	2.387	2.215	0.0851	0.1492
10	2.482	2.290	0.1150	0.1864
11	2.564	2.335	0.1448	0.2213
12	2.636	2.412	0.1738	0.2537
13	2.699	2.462	0.2016	0.2836
14	2.755	2.507	0.2280	0.3112
15	2.806	2.549	0.2530	0.3367
16	2.852	2.585	0.2767	0.3603
17	2.894	2.620	0.2990	0.3822
18	2.932	2.651	0.3200	0.4025
19	2.968	2.681	0.3398	0.4214
20	3.001	2.709	0.3585	0.4391
21	3.031	2.733	0.3761	0.4556
22	3.060	2.758	0.3927	0.4711
23	3.087	2.781	0.4085	0.4857
24	3.112	2.802	0.4234	0.4994
25	3.135	2.822	0.4376	0.5123
26	3.157	2.841	0.4510	0.5245
27	3.178	2.859	0.4638	0.5360
28	3.199	2.876	0.4759	0.5470
29	3.218	2.893	0.4875	0.5574
30	3.236	2.908	0.4985	0.5672
31	3.253	2.924	0.5091	0.5766
32	3.270	2.938	0.5192	0.5856
33	3.286	2.952	0.5288	0.5941
34	3.301	2.965	0.5381	0.6023
35	3.316	2.979	0.5469	0.6101
36	3.330	2.991	0.5554	0.6175
37	3.343	3.003	0.5636	0.6247
38	3.356	3.014	0.5714	0.6316
39	3.369	3.025	0.5789	0.6382
40	3.381	3.036	0.5862	0.6445

p: Number of laboratories.

TABLE A.13A

Parameters *h* and *k* for Mandel's Test for Significance Level $\alpha = 0.01$

		k								
						n				
p	*h*	2	3	4	5	6	7	8	9	10
3	1.15	1.71	1.64	1.58	1.53	1.49	1.46	1.43	1.41	1.39
4	1.49	1.91	1.77	1.67	1.60	1.55	1.51	1.48	1.45	1.43
5	1.72	2.05	1.85	1.73	1.65	1.59	1.55	1.51	1.48	1.46
6	1.87	2.14	1.90	1.77	1.68	1.62	1.57	1.53	1.50	1.47
7	1.98	2.20	1.94	1.79	1.70	1.63	1.58	1.54	1.51	1.48
8	2.06	2.25	1.97	1.81	1.71	1.65	1.59	1.55	1.52	1.49
9	2.13	2.29	1.99	1.82	1.73	1.66	1.60	1.56	1.53	1.50
10	2.18	2.32	2.00	1.84	1.74	1.66	1.61	1.57	1.53	1.50
11	2.22	2.34	2.01	1.85	1.74	1.67	1.62	1.57	1.54	1.51
12	2.25	2.36	2.02	1.85	1.75	1.68	1.62	1.58	1.54	1.51
13	2.27	2.38	2.03	1.86	1.76	1.68	1.63	1.58	1.55	1.52
14	2.30	2.39	2.04	1.87	1.76	1.69	1.63	1.58	1.55	1.52
15	2.32	2.41	2.05	1.87	1.76	1.69	1.63	1.59	1.55	1.52
16	2.33	2.42	2.05	1.88	1.77	1.69	1.63	1.59	1.55	1.52
17	2.35	2.44	2.06	1.88	1.77	1.69	1.64	1.59	1.55	1.52
18	2.36	2.44	2.06	1.88	1.77	1.70	1.64	1.59	1.56	1.52
19	2.37	2.44	2.07	1.89	1.78	1.70	1.64	1.59	1.56	1.53
20	2.39	2.45	2.07	1.89	1.78	1.70	1.64	1.60	1.56	1.53
21	2.39	2.46	2.07	1.89	1.78	1.70	1.64	1.60	1.56	1.53
22	2.40	2.46	2.08	1.90	1.78	1.70	1.65	1.60	1.56	1.53
23	2.41	2.47	2.08	1.90	1.78	1.71	1.65	1.60	1.56	1.53
24	2.42	2.47	2.08	1.90	1.79	1.71	1.65	1.60	1.56	1.53
25	2.42	2.47	2.08	1.90	1.79	1.71	1.65	1.60	1.56	1.53
26	2.43	2.48	2.09	1.90	1.79	1.71	1.65	1.60	1.56	1.53
27	2.44	2.48	2.09	1.90	1.79	1.71	1.65	1.60	1.56	1.53
28	2.44	2.49	2.09	1.91	1.79	1.71	1.65	1.60	1.57	1.53
29	2.45	2.49	2.09	1.91	1.79	1.71	1.65	1.60	1.57	1.53
30	2.45	2.49	2.10	1.91	1.79	1.71	1.65	1.61	1.57	1.53

p: Number of laboratories.
n: Number of results for one level.

TABLE A.13B
Parameters h and k for Mandel's Test for Significance Level $\alpha = 0.05$

		k								
		n								
p	h	2	3	4	5	6	7	8	9	10
3	1.15	1.65	1.53	1.45	1.40	1.37	1.34	1.32	1.30	1.29
4	1.42	1.76	1.59	1.50	1.44	1.40	1.37	1.35	1.33	1.31
5	1.57	1.81	1.62	1.53	1.46	1.42	1.39	1.36	1.34	1.32
6	1.66	1.85	1.64	1.54	1.48	1.43	1.40	1.37	1.35	1.33
7	1.71	1.87	1.66	1.55	1.49	1.44	1.41	1.38	1.36	1.34
8	1.75	1.88	1.67	1.56	1.50	1.45	1.41	1.38	1.36	1.34
9	1.78	1.90	1.68	1.57	1.50	1.45	1.42	1.39	1.36	1.35
10	1.80	1.90	1.68	1.57	1.50	1.46	1.42	1.39	1.37	1.35
11	1.82	1.91	1.69	1.58	1.51	1.46	1.42	1.39	1.37	1.35
12	1.83	1.92	1.69	1.58	1.51	1.46	1.42	1.40	1.37	1.35
13	1.84	1.92	1.69	1.58	1.51	1.46	1.43	1.40	1.37	1.35
14	1.85	1.92	1.70	1.59	1.52	1.47	1.43	1.40	1.37	1.35
15	1.86	1.93	1.70	1.59	1.52	1.47	1.43	1.40	1.38	1.36
16	1.86	1.93	1.70	1.59	1.52	1.47	1.43	1.40	1.38	1.36
17	1.87	1.93	1.70	1.59	1.52	1.47	1.43	1.40	1.38	1.36
18	1.88	1.93	1.71	1.59	1.52	1.47	1.43	1.40	1.38	1.36
19	1.88	1.93	1.71	1.59	1.52	1.47	1.43	1.40	1.38	1.36
20	1.89	1.94	1.71	1.59	1.52	1.47	1.43	1.40	1.38	1.36
21	1.89	1.94	1.71	1.60	1.52	1.47	1.44	1.41	1.38	1.36
22	1.89	1.94	1.71	1.60	1.52	1.47	1.44	1.41	1.38	1.36
23	1.90	1.94	1.71	1.60	1.53	1.47	1.44	1.41	1.38	1.36
24	1.90	1.94	1.71	1.60	1.53	1.48	1.44	1.41	1.38	1.36
25	1.90	1.94	1.71	1.60	1.53	1.48	1.44	1.41	1.38	1.36
26	1.90	1.94	1.71	1.60	1.53	1.48	1.44	1.41	1.38	1.36
27	1.91	1.94	1.71	1.60	1.53	1.48	1.44	1.41	1.38	1.36
28	1.91	1.94	1.71	1.60	1.53	1.48	1.44	1.41	1.38	1.36
29	1.91	1.94	1.72	1.60	1.53	1.48	1.44	1.41	1.38	1.36
30	1.91	1.94	1.72	1.60	1.53	1.48	1.44	1.41	1.38	1.36

p: Number of laboratories.
n: Number of results for one level.

TABLE A.14
Critical Values (λ_α) for Kolmogorov–Smirnov Test

α	λ_α
0.01	1.63
0.02	1.52
0.05	1.36
0.10	1.22
0.15	1.14
0.20	1.07
0.25	1.02
0.30	0.97
0.40	0.89
0.50	0.83
0.60	0.77
0.70	0.71
0.80	0.64
0.90	0.57
0.99	0.44

TABLE A.15
Critical Values of Regression Coefficient r_{crit}

f	$\alpha = 0.05$	$\alpha = 0.01$
5	0.75	0.87
6	0.71	0.83
7	0.67	0.80
8	0.63	0.77
9	0.60	0.74
10	0.58	0.71
12	0.53	0.66
14	0.50	0.62
16	0.47	0.59
18	0.44	0.56
20	0.42	0.54
25	0.38	0.49
30	0.35	0.45
40	0.30	0.39
50	0.27	0.35
60	0.25	0.33
80	0.22	0.28
100	0.20	0.25

Index

Page numbers followed by f and t indicate figures and tables, respectively.

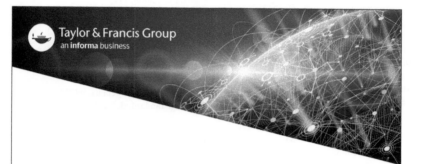